So Much Weather!

Facts, Phenomena, and Weather Lore from Atlantic Canada

by Gary L. Saunders

NIMBUS
PUBLISHING

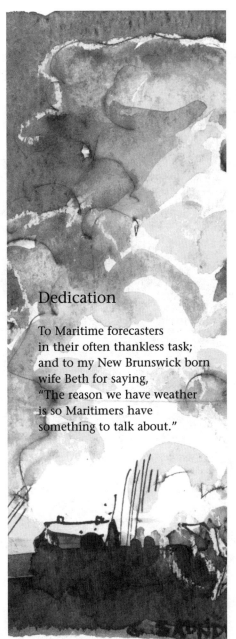

Dedication

To Maritime forecasters
in their often thankless task;
and to my New Brunswick born
wife Beth for saying,
"The reason we have weather
is so Maritimers have
something to talk about."

Nimbus Publishing Limited
PO Box 9166
Halifax, NS   B3K 5M8
(902) 455-4286

Printed and bound in Canada

Designer: Kathy Kaulbach, Paragon Design Group

National Library of Canada Cataloguing in Publication Data

Saunders, Gary L.
    So much weather! : facts, phenomena and
    weather lore from Atlantic Canada

    Includes bibliographical references.
    ISBN 1-55109-382-0

    1. Atlantic Provinces—Climate. I. Title.

QC985.5.A8S29 2001   511.65715
C2001-902221-2

Canadä

| The Canada Council for the Arts | Le Conseil des Arts du Canada |

We acknowledge the financial support of the Government of Canada through the Book Publishing Industry Development Program (BPIDP) and the Canada Council for our publishing activities.

Contents

Author's Preface

Sometimes on stormy afternoons when I was a boy, Grandmother Saunders would let me crank up her Victrola gramophone and listen to records. Lolling in her dim, cool, wallpapered parlour where the drapes stayed drawn except for weddings and funerals, I first heard the tinny faraway voices of Jimmy Rodgers, Woodie Guthrie, and Harry Lauder. It was magic.

There was one particular 78 I played over and over to hear "The Big Rock Candy Mountain." That song spoke to me. It wasn't just the candy, or the birds and the bees (which I didn't get anyway), or the cigarette trees (though secret smoking preoccupied me just then). It wasn't the lake of stew either, or the lemonade springs where the bluebird sings.

It was the whimsy of it all, the folly of working, the joys of being a hobo. Tramping the roads with all your worldly goods knotted in a bandanna on a stick! Riding the rails on fast freights, sleeping outdoors under the glittering stars every night, never washing your socks, wolfing down hot stew from a tin can, playing the guitar and yodelling like Jimmy....

Yes, I would run away soon. But much as I liked the song, its refrain puzzled me:

Oh, I'm bound to go where there ain't no snow,
Where the rain don't fall and the wind don't blow,
In the Big Rock Candy Mountain.

Why would anyone, especially a hobo, dislike wind, rain and snow? I certainly didn't, and I wouldn't when I grew up, either. I never understood why the adults in my life complained so much about the weather. I figured it was age and overwork. Whatever the sky was doing, especially if it stormed, was okay with me.

At 65 I haven't changed much. It surprises me when under-50s complain about the weather. Are we getting soft? I hope not, for if the weather pundits are right we're in for some hard times. Coddled by air-conditioned workplaces, indoor shopping malls and overnight snow clearance, we may need some of that childhood zest just to cope.

The weather *has* been woolly lately, even allowing for the media's obsession with bad news. Climatologists tell us the years since 1990 have been the warmest ever recorded. The resultant ups and downs of temperature, precipitation, and wind velocity have been extreme. While scientists disagree about what all this means—whether we're seeing climatic hiccups or a genuine atmospheric heart attack—my gut feeling is that things will get worse before they get better.

If so, the role of the meteorologist becomes more crucial than ever. Too bad, then, that our federal government has recently downsized its weather service. Some can see the loss in accuracy already. If ever people needed good advance warning, it's now—especially in these weatherful Maritimes.

In times like these it doesn't hurt to know a thing or two. Although my book celebrates our cantankerous weather, it also contains a lot of weather information—for instance, where we fit in the global picture, how weather works, and why the East Coast gets so damned much of it. It explores the ways in which weather touches our lives: landscape, plants and animals, folk-lore, language, health, even how we dress and house ourselves. Finally, the book probes thunderstorms, blizzards, and rare sky phenomena like sun dogs, haloes, and green rays. There's even a brief course in being one's own weatherperson, with tips on keeping a weather journal.

But I'm only an amateur. One thing I've learned in writing this is that meteorology is a vast, complex, and ever-changing science. A book like this can only skim the surface. For depth and detail, readers need to go to the professionals, some of whose works are listed starting on page 212. My intent is to recruit skywatchers, not to train experts. To me the sky is our crystal ball and nature's grandest spectacle. If we're in for worse weather, let's enjoy, not endure.

Foreword

When I was asked by Gary Saunders if I would write the foreword to his latest book, I jumped at the challenge. I especially welcomed the chance to pick up some of the "weather lore" that is such a part of life here in Atlantic Canada. Not only did I find some new lore, I also learned the reasoning or "story" behind much of it.

As Gary so adequately points out throughout the book, weather plays a part in just about every minute of everyone's life. This may be all the more so in our coastal communities, where the weather is at times so unpredictable and its power so devastating. As I read So Much Weather!, I came to realize that Gary did a lot more than research; he threaded the work with his own valuable knowledge and experience.

In my current profession as a television/radio meteorologist, I receive requests to explain a range of weather phenomena. To ascertain the most suitable answer, I consult many and varied books on the subject of weather, including books on such seemingly remote subjects as "word origins" and "how sayings originate." I found this latest work of Gary's to contain so much of everything—a real "one source document." Whether you're a teacher, a student, or an avid weather-watcher, this book delves into the whys and wherefores of weather, presenting factual, scientific meteorology in a simple, easy-to-read manner.

Peter Coade
Meteorologist
ATV/ASN/CTV
Halifax, Nova Scotia

The summer tourist in foggy Saint John, New Brunswick was getting exasperated with the wet weather. "Rain and fog, fog and rain, rain and fog," he growled. "When do you get summer here anyway?"

"Hard to say," smiled the hotel desk clerk. "Last year it came on a Friday I think."

People the world over jaw about their weather, but few do it more passionately than Maritimers. Why? The short answer is, because we get so much of it. "If you don't like the weather," we quip, "wait ten minutes and it'll change!" That's an exaggeration—but not by much. We do get more variable weather than almost anyone else in the so-called temperate latitudes.

People who say weather is boring can't be talking about *our* weather. Don't they know that TV weather channels now pull in more viewers than the news? That books and movies like *The Perfect Storm* are bestsellers? That "weather tourism" has suddenly become big business, with people spending thousands on "weather safaris" that chase tornados and thunderstorms across the American Midwest?

Although he was speaking of America, Mark Twain could have had the Maritimes in mind when he said, "There is a sumptuous variety about the...weather that compels the stranger's admiration—and regret. The weather is always doing something there;

A Most Weatherful Place

Into each life some rain must fall, especially if you leave the car windows open.
—High school student

always attending strictly to business; always getting up new designs and trying them on the people to see how they will go.... Yes, one of the brightest gems in the New England weather is the dazzling uncertainty of it." But then, Twain was living in Hartford, Connecticut, which isn't far away.

The story of our weather launches us on a journey from our own backyards to the ends of the earth and back, from prehistory to computer science and beyond. Of all the world's larger phenomena—earthquakes, volcanoes, tidal waves, avalanches—none is so pervasive and so powerful as the phenomenon we call weather. Weather touches each person on the planet every day and night of their lives. We swim in weather the way fish swim in the sea. The planet's last great wilderness is not down here—it's overhead.

Most of the time the weather is benign, for which we are thankful. When it turns ugly—which it has been doing more often of late—it can be as terrifying and destructive as any other force of nature. And we can't control it. That adds an element of fear. We may boast of deflecting incoming missiles or even asteroids with laser beams, but no reputable scientist seriously talks about deflecting hurricanes.

Ever-present, powerful, unpredictable, uncontrollable—the perfect evil giant. And Maritimers, like the boy in the fairy-tale *Jack and the Beanstalk*, live in the shadow of his castle. This also adds to the fascination.

And the giant delivers. Nova Scotia, Prince Edward Island, and New Brunswick get everything from near-tropic heat and brow-mopping humidity to bitterly Arctic cold that cripples cars and freezes plumbing. Between those extremes we get plenty of fog,

drizzle, and rain. We also get enough thunderstorms, blizzards, and hurricanes, as well as hail as thick as tennis balls, to keep us respectful. We even get thunder in winter. It's not unusual for temperatures to plunge from plus 10°C/50°F to minus 10°C/14°F in a few hours.

A single week can bring two or three major changes. In summer that means two or three days of sunny weather with westerly breezes, a day or two of cloud and mist from the east, and one or two days of rain from the south. In winter we can expect one or two days of cold, sparkling weather from the west, two or three days of cloud and flurries, and perhaps a two-day blizzard from the northeast.

Spring and fall conditions are even more variable. The weather can change two or three times a day. It's common to see several weather systems vying for dominance. One can read it in the clouds: puffy white fine-weather clouds to the north, with patches of blue, lambswool cirrus to the south (a warm front moving in), and low grey rain clouds giving intermittent showers. Our weather is among the most fickle in the world.

All this we take in stride. Some of us even like it. It's certainly never dull. A Nova Scotian friend of mine who spent a summer on the interior plateau of Mexico learned to appreciate Maritime weather. In that part of Mexico, he said, shaking his head in disbelief, every single day was sunny except for fifteen minutes of rain each afternoon, precisely at four o'clock. I wouldn't want to live there.

Here there's never a dull moment. Sure, a week of Fundy fog is boring while it lasts. But it doesn't last. And when the sun finally breaks through, how sweet it feels! When the fog overstays our patience, we try to remember what tourists say about Maritime summers and autumns: "The most pleasant on the North American continent!"

But we do take our weather personally. That's why we talk so much about it, why we make silly jokes about it. We even take a perverse pride in it.

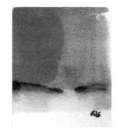

Let me here confess that some of my best conversations have opened with the weather. From weather, one can go anywhere. Like the air we breathe, it's one of the things every person on earth has in common.

Say you're in an elevator on a February afternoon and someone remarks, "Hear we're in for a blizzard." In no time at all you're talking power outages, or the price of fuel, or Rick Mercer's hilarious spoof about iced beer ads, or who's likely to win the Stanley Cup this year, or the curious marketing fact that

shoe stores can't sell winter boots until the season's first snowfall. By the time the elevator hits the main level, you both
feel sort of proud to be living in such a weatherful
place.

Anglers yarn about how they've never seen the
Miramichi or the Margaree or the Saint Marys River so
low as this year, and speculate on the effect on the salmon;
from there they discuss the need for stiffer quotas on the
high seas, to "what's the world coming to anyway?"

Gardeners and farmers talk weather at the drop of a
seed. House painters commiserate over how the summer fog and drizzle are reducing their income. School
boards agonize over whether to announce yet another
closure due to winter storms, and how they can ask teach-
ers and students to make up the time next spring.
Golfers...but enough.

About the only Maritimers who don't talk about the
weather are coal miners.

So let's admit our obsession. Let's not just talk about it, let's
start celebrating our weather. Fickle it may be, but that only ups
the ante. We profess to be nature lovers. Well, here on our
doorstep is nature's grandest spectacle, larger than any landscape, bigger than the sea, vast as the sky itself. It's the greatest
show on earth, a 24/7 wraparound extravaganza free for the
viewing, a spectacle that never repeats itself. Best of all, we don't
have to hunt for it, or book six months ahead. We know it's
coming. It comes to us. And we have a ringside seat.

Don't knock the weather, nine
out of ten people couldn't
start a conversation if it
didn't change once in a while.
—"Cirrus-ly Funny"
(Internet Weather Humour)

How Come *We* Get So Much Weather?

My Nova Scotia hiking buddies make fun of me because I always bring rain gear and a sweater. This comes from two summers I spent cruising timber on Newfoundland's Avalon Peninsula years ago. No matter how sunny the morning, it always seemed to drizzle or rain before we got back to camp. So I came prepared, and the habit stuck.

Not that the Avalon is notably wetter than certain parts of the Maritimes—for instance, Nova Scotia's eastern shore or the Bay of Fundy. It's not so much wetness as variability that prompts me to pack extra clothes. 'Fickle' is our weather's middle name. I wouldn't want to be a professional East Coast weather forecaster. Region for region, they have the hardest time of any in Canada. Why?

There are at least four reasons.

First, our region lies midway between the equator and the North Pole. We get weather from both directions.

Second, we live on the rim of a vast continent. Continents heat faster and cool faster than do oceans. That's why Toronto, in the same latitude as Halifax, is both hotter in summer and colder in winter. (Toronto's weather would be even more extreme without Lake Ontario's buffering effect.) The interiors of continents churn out cold, dry air in winter and hot, dry air in summer. Since

Sometimes out on the Bay of Fundy when the fog comes in thick, you can sit on the gunwale and lean your back up agin' it. So that's pretty thick fog out there. But you gotta be careful, 'cause if the fog lifts quick you'll fall overboard.

—Nova Scotia fisher

our prevailing winds are westerly, most of those air masses pass over us. The continual mixing of dry continental air and moist marine air generates much of our weather.

Third, we live on the shore of a vast ocean. Water, salt or fresh, has a remarkable capacity for absorbing and holding heat, so coastal waters take a long time to warm up and cool down. Whether air is warm or cool, moist or wet, depends on the temperature and humidity of whatever it touches. That's why all our easterly and southerly winds are warmer and moister than are our westerlies or northerlies. Beyond that, the sea cools us in summer and warms us in winter.

So our warmest weather comes after Midsummer's Day (June 21) and our coldest well after Christmas. June can bring hard frost, yet summery days can linger into November. Despite occasional incursions of frigid Arctic air, our winters are generally mild. Snowfall is abundant, but before it can accumulate, a warm interlude melts some or all of it.

Finally, two of the globe's great ocean currents meet and mingle off our coasts. Sail southeast from Halifax for half a day during any season and you will feel the air getting warmer. A day or two farther out, the colour of the water will change from greyish black to blue-green. You may see flying fish and sea turtles and other mementos of the south. Later, there may be floating

The sea is a heat sink. Just as it takes time to bring a kettle of cold water to the boil, it takes all summer for our coastal waters to warm up from the previous winter's chill. And just as the kettle of water stays warm for an hour or more afterward, so the ocean, once heated, holds the extra heat into autumn.

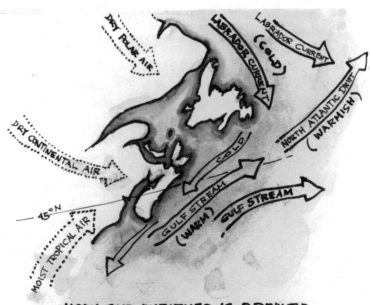

HOW OUR WEATHER IS BREWED...

skeins of Sargasso weed. You have entered the Gulf Stream, a warm current born in the shallow waters of the Gulf of Mexico and the West Indies under a subtropical sun.

But sail east, past Cape Race on Newfoundland's southeast corner to the Grand Banks, and you'll feel a steady cooling of the air. Not until you reach the mid-Atlantic will it lose its chill. That chill comes from the Labrador Current, which flows southward between Baffin Island and Greenland before veering toward Iceland. From December through June, ice pans and bergs ride this current south along the coasts of Labrador and Newfoundland until they melt on the Banks. Sometimes a very big berg runs aground.

At 45° north of the equator, central Nova Scotia is in the same latitude as southern France. If the Labrador Current were to veer east below Baffin, our climate could become much like North Carolina's—warmer year-round, less foggy, green Christmases, fiercer hurricanes.

On Newfoundland's northeast coast I have seen such bergs tower over the coastal cliffs, which were themselves over 30m/100ft high. In the early 1990s such a berg ran aground off the mouth of St. John's harbour in June. To the delight of tourists and photographers, it remained stuck all summer, streams of fresh water cascading down its glistening slopes. The berg was so huge that it made any easterly wind feel distinctly cool and moist—the Arctic on one's doorstep.

The mingling of these two vast ocean streams, one warm and one cold, creates frequent fog and rain along the Atlantic coast from eastern Newfoundland to Halifax. A smaller version of the same process occurs on the Isthmus of Chignecto, where cool Bay of Fundy waters squeeze moisture from warmer Northumberland Strait waters.

So there you have it. Moist ocean air from east and south mingles in all seasons with cool, dry, continental air from west and north, not only producing some of the most changeable weather in the country, but giving us the highest rate of precipitation east of coastal British Columbia. As a bonus, autumn hurricanes sometimes vent their dying rage on our boats and woodlands.

This continual atmospheric stir, pinwheeling and eddying across our rugged landscapes and seascapes like a giant river, creates the weather stew we are proud to call our own.

Maritime Weather: A Profile

Just how bad is our weather compared to that of other parts of Canada and the rest of the world? The answer is, not half as bad as we like to make out. We hold no world records for heat or cold or windiness. Variability is our thing.

In fact, when ranked with Canada's other regions we come off quite well. A lot depends on how you measure weather. Hottest single day? Coldest single night? Hottest August averaged over

CANADA'S 3rd WETTEST CITY

SYDNEY, NS

time? Greatest single snowfall? Greatest average snowfall by decade? "Normal" weather averaged over time?

In 1993 Environment Canada chose the latter method to assess the weather in each province and territory from 1961 to 1990. By collecting data from sixteen hundred weather stations located near seventy-five cities with a population of ten thousand or more, they came up with a result representing 70 percent of Canada's population. Then they chose twenty-five superlatives—"warmest summers," "highest snowfall," and so on—and compiled data for each. The results were intriguing.

Surprisingly, fifteen of the records went to centres outside Atlantic Canada. (However, we may have broken some local records for short periods; we're dealing here with regional averages).

Here are the Canada-wide results:

Warmest Summers: Kamloops, BC
Coldest Winters: Yellowknife, NWT
Warmest Places: Vancouver, BC
Lowest Annual Average Snowfall: Victoria, BC
Most Days Below Freezing: Thompson, MB
Fewest Days Below Freezing: Vancouver, BC
Longest Frost-Free Period: Vancouver, BC
Shortest Frost-Free Period: Thompson, MB
Driest Cities: Medicine Hat, AB
Most Thunderstorms: London, ON
Sunshine Capitals: Estevan, AB
Sunniest Summers: Yellowknife, NWT
Fog-Free Cities: Penticton, BC
Most Humid in Summer: Windsor, ON
Clearest Skies: Estevan, AB

The Maritimes' BEST WEATHER

P.E.I.

The good/bad news is that Atlantic Canada at least *placed* in six categories. (Note: Since Maritime weather is a regional affair, I've included Newfoundland and Labrador in the picture.)

For **Wettest Cities**, St. John's came second, followed by Sydney, Halifax, Saint John, Moncton, and Corner Brook. The winner was Prince Rupert, BC, with 2,552mm/100in of precipitation per annum. Under **Fewest Thunderstorms per Year**, St. John's ranked third after Victoria (which had three), while Corner Brook and Sept-Iles trailed in seventh and eighth positions. For **Sunniest Winters**, Saint John and Halifax earned seventh and eighth after Winnipeg (358 hours). In **Number of**

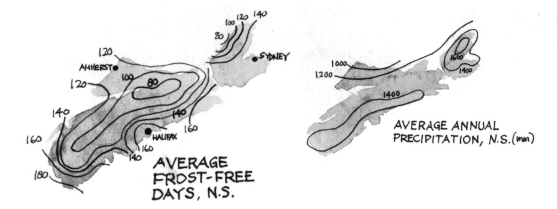

AVERAGE FROST-FREE DAYS, N.S.

AVERAGE ANNUAL PRECIPITATION, N.S. (mm)

Days with Blowing Snow, St. John's and Charlottetown got fourth and fifth after Chicoutimi (37). In **Cloudiest Skies**, St. John's came second after Prince Rupert (6,123 hours), with Sydney and Halifax trailing in seventh and eighth place.

We came first in two classes only, namely (can you guess?) **Foggiest Cities** (St. John's at 121 days, just ahead of Saint John, Halifax, and Sydney, which were trailed by Moncton and Sept-Iles); and **Glaze Capitals** (St. John's, with 38 days of freezing precipitation, followed by Sydney in third, Charlottetown in fifth, and Halifax in ninth).

We did set some mini-records within the region. In August 1974, the community of Northwest River in Labrador recorded a temperature of 41.7°C/107°F. In 1973, Esker 2, a station in central Labrador, recorded a low of -51.1°C. In 1977-78, Woody Point on Newfoundland's west coast got 893cm/35in of snow. For hours of sunshine, Halifax topped the region with 2,246 in 1978. Newfoundland's Belle Isle and Halifax nearly tied for maximum sustained hourly winds, clocking gusts of 138kmh /83mph and 135kmh/81mph. For maximum precipitation, Wreck Cove Brook in eastern Cape Breton topped the region at 2,360mm/93in—twice that of New Brunswick. Weatherly speaking, the region's most benign province is Prince Edward Island.

None of which should surprise us. We do have a climate of extremes—even though our extremes are less extreme than those of other parts of Canada and of the world. Clearly, for its size our region sees more diversity of weather on average than almost any other part of the country.

Was the same true in our parents' and grandparents' time? Let's have a look.

CANADA'S 2nd FOGGIEST CITY
SAINT JOHN, NB

For some reason, in the winter of 2000-2001 St. John's was fog-free for three consecutive months, an unheard-of occurrence.

3

Memorable

East Coast

Weather

In the mid-1980s, I was attending a morning forestry meeting in an office building near the Halifax waterfront. After the meeting, several of us took the elevator to the top floor, where there was a cafeteria with a fine view of the harbour. It was an overcast day, so I was surprised to see the place lit by an eerie white glow. Walking over to the windows on the water side, we saw the harbour filling with drift ice. The ice had been reported off Sydney some days before, and now, nudged by northeasterly winds, sealed off the eastern shore as far west as Halifax.

While this was unusual, it was not unheard-of. Most weather events are like that. They may startle us, they may seem unprecedented at the time, but that's because weather moves in much longer cycles than we do. All of the storms mentioned in this chapter are of that sort. At the time they may have seemed like harbingers of doom, but in the long view they were simply climatic aberrations.

"There was one winter," said an old itinerant Maritime preacher, "so monstrous cold people had to resort to desperate measures. I know, because I often stayed overnight at parishioners' homes. After one particularly cold night, during which I shivered under my scanty bedclothes, I came downstairs to breakfast. Seeing me on deck, the good woman of the house called to one of her girls, 'Marie, go up to my bed and get a loaf of bread for the minister's breakfast.'

"Is that," he concluded, "where the term *Bed & Breakfast* came from?"

Anyone with older relatives has heard at least one story of this sort. While it's true that memory will embellish and that snowdrifts look higher to children, not all such tales are tall.

Take snow, for instance. ("Take it all," says someone over forty.) Since our ancestors lacked mechanical plows, weeks might pass after a major storm before shovel brigades managed

Climate is what we expect;
Weather is what we get.
—High school student

to clear the roads and bridges linking towns and villages. Trains were forced to wait for days while work crews or steam plows cleared the track. My parents once spent three days and two nights stranded on a snowbound train. Resourceful porters shifted the relatively few passengers into three or four cars where heat, water, food and blankets could more easily be supplied. A woman had a baby while the train idled.

In the woods, blizzards caused even more trouble. While lumbermen needed snow for hauling—indeed they pulled special water-sprinkling sleds over the snow at night to make it more slippery—new snowfalls forced them to break trail before oxen or horses could get through. Also problematic was trying to fell trees in deep snow. The men couldn't get at a tree's base without shovelling, and shovelling reduced output. So they simply cut the stump high, wasting good wood in the process. Even so, when the tree began to fall, the logger had no room to escape if it twisted or fell the wrong way.

Snowstorms weren't the only villains in oldtime weather tales. The Great Ice Storm of 1998 wasn't all that long ago, and it wasn't a Maritime storm. Yet no Maritimer will soon forget the most destructive winter storm in our Canadian history. For several weeks, Atlantic Canada and New England were part of a massive effort to house, feed, and provide light to the homeless, sending truckloads of food and other emergency supplies to southern Quebec and eastern Ontario. Over eighty hours of non-stop freezing rain and drizzle freighted power poles and hydro lines with up to 8cm/3in of ice. Giant steel pylons crumpled like wet cardboard. The storm left four million people in eastern Ontario and southern Quebec freezing in the dark for up to three weeks. Montreal and other cities were without power for upwards of two weeks. Three weeks into the relief effort, seven hundred thousand people were still in the dark and cold. Farms and factories that lacked backup generators were forced to shut down, and hundreds of baby animals perished. Schools had to close. Gyms and churches were packed with evacuees. A tenth of urban trees were toppled or mangled, and Quebec's maple syrup industry, the world's largest, was badly mauled.

Fortunately for the Maritimes, the ice storm barely brushed us in passing. Down here it behaved more like a typical January thaw. The unusual surge of polar air that shoved the warm tropical low so far inland just didn't deliver enough of the supercooled (i.e., below freezing) water that instantly freezes on contact with pavement, gravel, tree branches, and wires. Moreover, we didn't get Montreal's atmospheric inversion—a layer of

Jan. 11

Light snow on a SE wind, then freezing rain coating everything. Driving is treacherous. The temp. rises until the ice that drags down the power lines and bends the trees like old men has melted and they both spring back. Where they don't, there are power outages, branches are stripped, birches bow down left and right as in Frost's poem. In Newfoundland this is called a "glitter storm," in the Maritimes a "silver thaw".

—Author's journal

Overnight, ice encased every twig, cone, bud, branch, and trunk for tens of miles in every direction... Glassy cylinders hugged every wire, fencepost, and clothesline pole. Doorsteps, woodpiles, rocks, pebbles, cars, trucks, tractors, roadside garbage drums, and letter-boxes— they all shone under this new, hard, gleaming, and transparent skin.

—Harry Bruce, Down Home: Notes of a Maritime Son

warm, moist air trapped over a stable layer of frigid air. The cold lower layer squeezed moisture from the warm upper layer, and the descending water froze while remaining liquid. This happens when chilled droplets contain too few "freezing nuclei"— microscopic clay and other particles that resemble six-sided ice crystals and so trigger ice formation. All we got was some ice buildup at higher elevations.

That ice storm forcefully reminded easterners of what several days of freezing drizzle can do. In a few hours it destroyed a power grid that took a generation to build and that was considered impregnable. Parts of it had to be rebuilt, not repaired. It forced utilities throughout the northeast to take a hard look at their equipment and their suppositions. Old weather stories like the following suddenly took on new meaning.

Nova Scotia

Arguably the stormiest province in Canada, this 600km/400mi long peninsula juts almost into the path of autumn hurricanes whirling up from the Caribbean and the Gulf of Mexico. So it is regularly raked by gales and worse. David G. Dwyer, a forester who did his 1965 bachelor of science thesis on the history of forest blowdowns in his native province, discovered a fairly cyclic pattern to such winds. By quizzing local oldtimers and studying tree rings going back nearly 400 years, Dwyer found that big blows occurred every few decades, with a monster wind virtually guaranteed every 100 to 130 years.

He found support for his theory in the logbooks of Titus Smith Jr., the young Halifax naturalist who surveyed mainland Nova Scotia's forests for Governor John Wentworth in the summers of 1801 and 1802. Almost everywhere Smith went, he found widespread wind damage, both recent and from decades ago. Sometimes the trees were so tangled he had to clamber over them or crawl under them. Dwyer concluded that Nova Scotia's woods had suffered major windstorms around 1650, between 1700 and 1710, in 1790, 1820, and between 1869 and 1880. The same winds would have severely damaged the forests of New Brunswick, Maine, and PEI.

The coal mining town of Glace Bay, NS got its name (French = _glacé_) from the pack ice that still blockades this coast in late winter. The French mined coal at nearby Port Morien to heat Fortress Louisbourg, where sizable trees were rare.

Though Bluenosers suffer a lot of wind, they seldom see drift ice except in Cape Breton. Residents of Sydney have waked to see the harbour full of pack ice more than once. In June of 1970, Neil Van Nostrand and I were icebound on nearby Scatarie Island for nearly ten days. Our work with a Newfoundland-to-Nova Scotia ptarmigan transplant finished, we could only beachcomb and wait. We stopped shaving. The ice finally pulled out on a westerly wind; I kept the beard as a memento.

E.M. Deyarmond's *The Whip-Handle Tree*, an informal history of central Nova Scotia's Stewiacke Valley, lists several other disastrous storms—some local, some province-wide. The "great freshet" of 1792 washed the stooked wheat downriver and stranded many families. The "Big Wind" of November 12, 1813 flattened large tracts of woodland and demolished many homes and barns.

People called 1816 "The Year Without a Summer." The Indonesian volcano Tamboro had erupted the year before, spewing a cloud of dust and ash and sulfur into the stratosphere. The heavier dust and ash particles settled out in a few months, but the sulfur molecules bonded with water vapour to form feather-light aerosols that circled the globe for years. The eruption killed fifty thousand people outright, but thousands more died from crop failures due to atmospheric cooling.

Snow fell somewhere in the northeast every month of that year. Similar weather prevailed in upper latitudes around the world. In the Maritimes, frost struck in June, July, and August. In September a killer freeze blackened any crops still unscathed. Men worked the fields wearing mittens. At the Fourth of July celebrations on the green in Woodbury, Connecticut, everyone wore overcoats and the men pitched quoits with gloved hands lest their fingers stick to the frosty iron.

One of the Maritimes' worst storms was the Saxby Gale (or Saxby Tide) of October 5, 1869. It was so called because in November 1868 Mr. S.M. Saxby, a civilian instructor in the British Navy, published a letter in the London press warning of various cataclysms a year hence. He based this prediction on an unusually precise alignment of sun, moon, and earth to occur at 7:00 A.M., October 5 the following year.

Mr. Saxby likely knew of the ancient Chaldean discovery that every 18 years and 11.3 days (10.3 if the cycle includes five leap years) the sun, earth, and moon align in a certain way—the so-called Saros Cycle. The great storms of 1705, 1759, and 1869 all occurred during those alignments.

Right on cue, on October 3 a tropical storm brushed Washington, Philadelphia and parts of New England. "By the evening of October 4," wrote the late Dr. Albert Roland in his *Geological Background and Physiology of Nova Scotia*, "waves were crashing over the wharves of Annapolis and Saint John, and shingles were being torn from the roofs by the hurricane winds. Over eighty buildings were destroyed on Campobello Island, and abnormally high tides swept up the Bay of Fundy...."

On the night of June 6th, water froze an inch thick, and on the night of the 7th and the morning of the 8th, a kind of sleet or exceedingly cold snow fell, attended by a high wind, and measured in places where it drifted 18 to 20 inches in depth.

—Danville, Illinois North Star

As the tide roared eastward up the narrowing funnel of Minas Basin, it overwhelmed dykes which had stood since Acadian times. The mile-long Wellington Dyke on the Canard River—15m/50ft high, 37m/120ft wide at its base, seven years in building—which had stood intact since its completion in 1823, was

breached. Across the Cornwallis River estuary at Grand Pré it was just as bad:

"At 11 o'clock Monday night," reported the *Halifax Chronicle* on October 8, "four dykes...gave way, and ten minutes afterwards the lowlands for miles around were flooded.... Cattle...drifted out to sea."

There was no way to measure the winds, but Dwyer declared that "tides in the Bay of Fundy rose above any mark ever before recorded."

At the head of Chignecto Bay near Amherst, a 21.6m/71ft head of water—nearly twice the norm for that bay—overwhelmed the Tantramar dykes. On the New Brunswick side of the narrow isthmus, roaring waves built a sandbar off Sackville's river channel. The channel has been plugged ever since. Moncton on the Petitcodiac River to the west saw a 2m/6ft surge that broached several dykes.

Other harbours were silted up that night. Before the storm, the wooded basalt dome of Partridge Island near Minas Basin's north shore had been a haven for sailing ships seeking shelter on this open coast. Though connected to shore at low tide, it offered good anchorage and deep water east and west of the bar—which is why the island was settled before Parrsboro.

But due south looms Cape Blomidon, its curving red sandstone bluff pinching the 18km/11mi-wide basin down to a mere 10km/6mi. Roaring through this bottleneck, the monstrous tide scooped boxcar-loads of cobbles and sand from the windward beaches and dragged them along shore until they hit the island. What had been a low sandbar soon became a wedge of new land, half a kilometre to a side, joining the island to the mainland by two curving ridges. Today the saltmarsh in the middle gets flooded only by the highest tides.

That same night, in Parrsboro Harbour a few kilometres north, a large square-rigger was torn from its moorings and flung up onto the river bank, but suffered little damage. As the tide was funnelled into the shallow wedge of Cobequid Bay, it mounted higher and higher. More dykelands were flooded, and though Truro at the bay's head had no shipping facilities, the storm did enough damage for a road (Saxby Lane) to be named after it.

Meanwhile, Cumberland Basin to the west was being hammered. In the Nappan area south of Amherst, a tidal surge uprooted barns and drowned cows, sheep, and pigs. By morning every cove was strewn with haystacks and dead animals. It took days to sort out who owned what hay.

As the storm increased, almost everyone fled to high ground. The *Amherst Gazette* of October 18, 1869 tells of four men who had gone to Fort Lawrence Creek earlier to secure a schooner:

They had to seek shelter in a barn. The tide rising, they abandoned the barn and took to a fence [extending] to the upland, and by passing along which they hoped to be safe. The waves swept away the fence. Two...managed to reach some poles and save themselves. The others...were drowned.

On August 25, 1873 the Great Nova Scotia Cyclone struck, bringing gale-force winds to Halifax, Truro, and Sydney. Besides taking five hundred lives, it destroyed twelve hundred vessels, nine hundred buildings, and a great number of wharves, dykes, and bridges. Property losses were conservatively estimated at $3.5 million—nearly $80 million in today's dollars. A Sydney weather observer called it the worse gale since 1810. And because Halifax lost power, they had no means of warning the eastern counties. It helped convince governments that Canada needed a better storm warning system.

Contrast that with today's use of Doppler radar, which measures the speed and direction of approaching fronts and winds, enabling forecasters to give more accurate advance warning.

The Year of the Big Snow

Oldtime Bluenosers named 1905 "The Year of the Big Snow"—though more than one storm was involved. Truro, located near the centre of the 400mi/600km peninsula and the natural hub of the province's railway network, suffered the worst disruption.

According to Truro's *Daily News*, the snows of 1905 came in several batches from early January through late February. While the storms were bad for business, they were good for copy. The January 26 edition of the *Daily News* carried this item:

We have had storms and cold, ad libitum, all over the country this winter, but the effort last night was a supreme one. The "old fashioned" winter is not asked for any longer, it is right here in our midst.... Trains are stalled in every direction, and the country roads are almost impassable.

A day later it reported:

Number 10 train from North became stuck on the 25th in heavy snow over 20 feet deep between Londonderry and East Mines. Some 50 men with two engines were working all night to get this train out of her trouble.

Four days later the editor complained:

When can the cold and snow cease? The "old-fashioned winter" man has taken a back seat, and well he might, as during all winters that has helped frost his venerable locks, he can recall none more severe than this of 1904-05.

Even in ordinary winters Truro usually gets more snow than anywhere else on mainland Nova Scotia, for two reasons. It sits at the head of the southeastern arm of the Bay of Fundy. This tide-churned, wind-lashed arm of the Gulf of Maine funnels cold dry westerlies directly into central Nova Scotia and southern New Brunswick. On the way, the winds pick up moisture from the bay, whose waters seldom drop below 10°C/50°F even in winter. When someone suggested that God was punishing certain goings-on in the town, the editor intoned:

Friend, perhaps they're trying to bury up sin, of so horrid men, that to be hated needs but to be seen, with piles of pure white snow.

On February 6 a woman reminded the editor about the old superstition that the weather on Candlemas Day (February 2, a church feast with candles commemorating the purification of the Virgin Mary), foretells the weather for the rest of the winter. (Our Groundhog Day comes from this.) The editor noted that the "...day was fine, crisp and cold—a prognostication that we are to have very severe weather for the rest of the winter."

Three days later, the winter's worst snowstorm walloped the region:

The present snow storm on the ICR [Intercolonial Railway, Halifax-Montreal] is perhaps the greatest in the history of the road. In 1872 there was a bigger and longer snow block, especially in the Eastern Division than at the present, but the appliances for fighting snow and ice could not compare to the big engines and modern snow ploughs everywhere around us today. We are pretty safe saying the ICR has just faced its heaviest snow block....

It was to be the first of several massive snowfalls that month. Reports of stuck trains came in daily from across the county. Our editor, hard put to find adjectives potent enough to describe the seige, outdid himself:

The blizzardous [sic] wind that swept the country on Saturday 11 completely blocked traffic with heavy snow in the ICR between Halifax and Shubenacadie as never before experienced in the history of the road. Blocked in the deep drifting snow between the two stations, from noon Saturday until ten o'clock Sunday morning there were six passenger trains with fourteen engines.... The train men, after working for hours trying to dig their engines and cars out of the snow, finding their efforts unsuccessful, sent word to Truro for assistance.... During the day the other trains which had been tied up at Shubenacadie, Enfield, Wellington and Rockingham, got clear of the great snowbanks which had held them for 20 hours in their grip.... The average height...through which the trains plowed their way was about seven feet. The train which left Windsor that day [on the Dominion Atlantic line] got as far as Kennetcook where she has been stalled. A few days ago an engine and auxiliary car with a crew of men started to assist in digging out the helpless train. This engine and car only got out from Truro a mile and a half, when they got stuck. All further efforts to clear the road until after a change in the weather comes, have been abandoned.

As the flow of goods and services dried up, people began to feel the pinch. Monday, February 13:

Truro is threatened with a coal famine as it has been impossible to get coal from our mines over the ICR...if this snow blockade continues...the inevitable will happen and we will be without electric light, power or heat.

Three days later:

The present storm, coming on all that has been, has blocked rail transportation beyond anything heretofore experienced. There has not been a train in Truro today and the only train leaving this station was the early morning local which is now stalled at the Elmsdale bridge. No. 10 train is stuck at Wentworth and No. 76 is somewhere on the Mountain. Robert Lightbody, one of the veteran drivers on the ICR, says in all his experience he has never went [sic] through anything equal to the blizzard of February 10-11. Driver Lightbody says the train crews were completely exhausted by the work and the drivers were worn out and wet to the skin with snow water and their clothes were frozen on them.

Things got worse. The Saturday, February 18 edition reported that Truro hadn't seen a train for days. Monstrous drifts on the other side of Mount Thom had cut them off from the north and east.

On the line to New Glasgow...it is almost impossible to handle the great quantities of snow encountered.... Last night a big snow plough was pushed through the drifts to New Glasgow. This morning the plough was started back again for Truro, but was, at the time of going to press, stuck in a mountain of snow near Glengarry. No train has arrived in Truro since Thursday night the 16th.

Finally, on February 27, the snow let up a bit.

The rain that has been falling lightly since 10 o'clock last night, is settling the snow and assisting in clearing the rails all around, and so long as the temperature does not take a drop, is one of the most favourable assistants the railways can have.

Seizing the opportunity, work crews toiled round the clock to free trapped engines. Since snowplows can't butt through drifts much deeper than themselves, the men had to use shovels:

Passengers were practically buried in the snow less than two miles north of Londonderry. At midnight last night the men were still six telegraph poles this side of the snowbound express, shov-elling through a bank 22 feet deep. The cars for the most part were completely buried and the tops of the smoke stacks were the only parts of the iron horses showing above the snow.

That's the kind of winter oldtimers bragged about. Yet despite the near collapse of civilization, only one hockey game was can-celled in Truro that month.

In September 1942 a cloudburst after two days of continuous rain burst several dams on the Stewiacke and its branches, tak-ing farm wagons and machinery downriver. Two loggers drowned when their camp washed away. Terrified horses and sheep swam and sank among tumbling, uprooted trees. A hen house occupied by 350 screaming birds whirled away on the flood.

A tornado in June 1949 unroofed two buildings and toppled trees on Harry Fulton's farm. That September a sudden hail-storm fired stones as big as hen's eggs through windows in the Stewiacke Valley, stripped fruit trees, and forced anyone caught outdoors to lie face down. At noon the next day unmelted chunks of ice still littered the ground.

Three hurricanes of the mid-fifties, Carol, Hazel, and Edna— at that time all hurricanes north of the equator were given female names and those south of the equator male names— wreaked havoc in eastern Canada's woodlands, uprooting and splintering old conifers across thousands of hectares. Edna (September 1954) was followed by a massive winter ice storm. Hazel did most of its damage in Ontario, but its fury extended into the Atlantic provinces. Overall it killed eighty people and caused $100 million worth of damage.

In Nova Scotia alone, these three windstorms demolished the equivalent of a year's supply of wood. Since the province then had only one newsprint mill, no ready market existed for so much low-grade wood. It had to be left on the ground for insects and fungi to consume. But most of the sawlog material was sal-vaged via emergency leases and contracts with sawmill opera-tors.

In September 1997 Puerto Rico imported 50 tonnes of Nova Scotia snow in time for Christmas. The snow was shipped 3,000km/1,800mi on a refrigerated container from Ski Martock near Windsor.

New Brunswick Storms

With a vast continent at its back, this province has been called the least maritime of the Maritimes. The weather is also less capricious. Except along its Fundy shore, winters tend to be colder and summers hotter here, especially west of the Saint John River. While the province is not the wettest of the Maritime provinces, it is the snowiest. That's because its land mass extends farther north, and its western half is mostly high ground. Not only does more snow fall, it tends to stay.

This was brought home to me while attending university in Fredericton in the 1950s. A blizzard would sweep up the Saint John Valley and last for two or three days. Then the sun would come out, and you could count on a week or even two of fair weather. Morning after morning, walking to classes from my boarding house high on Hanson Street, I could hear the squeak of my boots in powdery snow. Often the temperature would hover around -23-29°C/-10-20°F. One morning the mercury stood at -41°F/-41°C. Yet it didn't feel that cold, because the air was dry. It numbed the fingers and nipped the nose and burned the lungs, but it didn't penetrate one's clothing much.

With each storm and flurry, the snow piled up. A mild spell would settle it, but the top layers insulated the bottom layers, so most of it stayed. By April the snowplow crews had no place to put any more. And when it melted, there was water everywhere. What with runoff, and ice jams along the river, the downtown often flooded. One such flood left watermarks a good metre above the sidewalks, which wasn't surprising, since the city's whole downtown sits on a loop of the floodplain.

New Brunswick's snowiness became even more obvious after we moved from New Brunswick to Nova Scotia. My wife had family north of Moncton, and every Christmas we would take the children for a week-long visit. The difference in snow accumulation was obvious. Nearly always they had more. The kids were delighted. At Grammy's they could toboggan and build snow people and have snowball fights when they couldn't at home, 3.5 hours' drive away. It wasn't that Nova Scotia didn't get snow, it did; but the snow rarely lasted more than a few weeks before a thaw took it away.

Of course New Brunswick gets its share of fall hurricanes. The Saxby Gale of 1869 was essentially a hurricane. It destroyed wharves in fishing communities on Grand Manan, wrecked boats in Saint John, and drowned much livestock. Other hurricanes have wreaked havoc with the province's mainly coniferous forests. Generally, though, the province fares better than its two Maritime neighbours.

Hurricane Hazel (October 15, 1954) dumped an estimated 300 million tonnes of rain on Toronto.

Flooding is the real nemesis, especially for those who live along the lower reaches of major streams like the 640km/400mi Saint John. One reason is the province's snowfall accumulation. This, combined with having the Maritimes' highest mountains (Mount Carlton 820m/2690ft), and its longest river, is a recipe for trouble.

During the twentieth century alone, the lower Saint John valley has seen at least six floods costing one million dollars or more. The worst happened during the last week of April 1973, when downtown Fredericton had 260 homes and 193 businesses awash and the farms upstream and down were lakes. Total damage to buildings, farm machinery, and livestock was valued at $11.9 million—over $45 million in today's dollars.

Two things brought on the flood of 1973. First, the watershed's snowpack was unusually deep, in places up to three metres, or nearly ten feet. Three metres of snow melts down to 30cm/12in of water. Second, on April 27, two days after most rivers had already crested at above normal levels, a rainstorm swept northern and central New Brunswick. It dumped 7.5cm/3in of water over the snowpack and ushered in a sudden thaw. The rest, as they say, is history. They also say that such floods occur here roughly once in seventy years. Hmmm...2043. Frederictonians could rest easier if they knew that for sure.

Saint Johners would rest easier too, if they could be lighthearted about February 2 like other Canadians. Those who recall the Groundhog Day Storm of 1976 cannot. On that day, one of the fiercest gales ever to hit the Bay of Fundy descended on southern New Brunswick and the surrounding region. At its peak, waves reached heights of 12m/39ft in the harbour and winds were clocked at 188km/113mi per hour. It was said that no storm since the Saxby Gale had done so much damage to the area. Besides knocking out power and closing most highways for a week, the blizzard wrecked buildings and docks, battered mobile homes and vessels, and tore away huge chunks of shoreline. For miles inland, trees and rocks were coated with sea salt from windblown spray. To add to the misery, frigid Arctic air poured in behind the gale, freezing and splitting water mains.

Prince Edward Island Storms

"A moist and temperate isle" is how Canada's meteorological service describes it. It's hard for mainlanders to picture this province as anything but peaceful. Winter is mild, spring is late but worth waiting for, summer is "modest and breezy," fall is stunning. Hurricane Alley seems far away. Yet the ingredients are there. Dry continental air to the west, moist ocean air to the east; the warmest summertime waters north of Carolina lap its Northumberland and western shore; frigid Labrador currents swirl in from the Gulf of St. Lawrence. It is inevitable that every now and then a big one will come along.

Such a storm blew up on February 21, 1982. Though the winds seldom exceeded 100kmh/60mph, they lasted a week. And though the 50cm/20in of snow that fell was not a huge amount, the island's gentle topography allowed the gales to shove the snow around. Drifts as high as 7m/25ft engulfed two freight trains and even forced snowplows off the Trans Canada Highway. No sooner had the great yellow machines plowed open a lane than it closed again. The wind chill was so low that going outside was dangerous. Toward the end, government helicopters had to ferry food, fuel and medical supplies to marooned residents. No human lives were lost, but it took two weeks to put things back to normal.

Meanwhile, across the Cabot Strait...

What happens to Maritime weather when we're done with it? We pass most of it on to Newfoundland and Labrador. Generally, storms track northeast, picking up lost moisture as they cross the 100km/60mi water gap. Let's see what our neighbours have to put up with.

As a child, I often heard oldtimers speak of the "Seventh of April Batch" that buried central Newfoundland in 1907. How it compared with the Great Snow of 1872, or the monster snows of 1904 and 1905, I don't know. Unlike some tall weather tales, its details ring true.

My father told me, and his father told him, that the snow was so deep a horse belonging to the Horwood Lumber Company perished in its log stable before teamsters could dig it out. The drifts were so deep along the company's 8km/5mi portage road that it took five days of hard shovelling to reach my home community, Clarkes Head in Gander Bay. He told me that one of the shovellers laid a plank across the diggings and stood on it as huge Nova Scotia work horses walked under it without touching their harness hames. A full-grown Clydesdale stands nearly 2m/6ft at the shoulder and the hames would add nearly 0.3m/1ft.

Over on the south side of the bay, a Mrs. Hodder tied her garter to the tip of a young fir tree as a mark. The next summer she delighted in showing it to incredulous visitors. People with outdoor firewood piles had trouble both getting to them and digging out the buried chunks.

Though my father didn't recall that snowfall—he was only three at the time—he remembered another that reached "nearly to the top of a telegraph pole." Telegraph poles in rural Newfoundland were shorter than today's, but not much shorter than 7m/23ft. So the snow had to have been at least 5m/16ft deep.

Labrador too has had its share of fierce storms. One storm in 1855 destroyed three hundred schooners, took three hundred lives and left two thousand people destitute. The sinking of the *Caledonia* in 1875 left 82 men, women, and children stranded on an iceberg all night. Similar incidents peppered the 1920s, a time when hundreds of Newfoundland schooner captains took their families north for the summer to fish cod for West Indies and European markets. Sailing home in the fall, many schooners would be loaded to the scuppers and would not answer the helm too well. They were easy prey for freak winter storms, fog and ice. Hundreds of people never made it home, let alone profited from their summer's toil. Before the Labrador fishery ceased in the fifties, scarcely a cemetery in northern Newfoundland lacked such victims. Like the victims of the sealing disasters of the late nineteenth and early twentieth centuries, they were human sacrifices in a deadly cat-and-mouse game with the elements, with the prize being a living wage.

What many consider the worst blizzard ever to hit the island of Newfoundland struck on February 16, 1959. Though it killed only six people, it left seventy thousand people without services and blocked roads and rail lines with drifts up to 5m/16ft deep.

By a quirk of nature, another infamous storm struck the area on nearly the same date. The storm was called the "Ocean Ranger Storm," after it sank the Ocean Ranger oil drilling platform. The rig sat 300km/180mi east of St. John's. On Monday February 15, 1982, a fatal combination of high winds, massive waves, and human error took the lives of 84 crewmen. Like the *Titanic*, the rig seemed unsinkable; but its water ballast was put in the wrong tanks, and mountainous seas did the rest.

That killer gale began as a whirl of warm moist air in the Gulf of Mexico three days earlier. By the time it reached the Grand Banks it was packing winds of up to 168kmh/100mph and was raising waves as high as five-story buildings. The rig, battered for most of Sunday, developed a 12- to 15-degree list to port. In the pre-dawn hours of Monday, orders were given to abandon ship. Around daybreak the rig went under. Search and rescue teams, hampered by snow, freezing rain and spray, did their best but

The "Independence Hurricane" that hit Newfoundland on September 9, 1775 drowned several thousand British seamen.

found no survivors. Next to the North Sea tragedy of March 1980, which took 123 lives, it was the world's worst offshore drilling disaster. The same gale sank the Soviet container ship *Mekhanik Tarasov* 120km/72mi to the east, drowning 33 seamen.

A Note on Heat Waves

"It is so hot in some places," wrote a young student, "the people there have to live in other places." Maritimers are fortunate not to be among those. In fact, people come here in summer to cool off. Heat stroke and related illnesses are not the problem for

us that they are for people farther west and south. Maritimers by definition live too near the sea to suffer the scorching days and suffocating nights so dreaded by Torontonians and New Yorkers.

Sure, we gripe a bit when temperatures rise above 30°C/86°F, but our grumbling is without conviction. For we remember winter. In fact, a Martian tuning in on a Maritime forecast would think our God was sunshine and our Devil rain. We've become sun worshippers like everyone else.

Yet, give us two or three weeks of uninterrupted summer sunshine, and we droop like unwatered geraniums. Give us four weeks, and we're grumping about parched sinuses, or the need for liquids and salt and a trip to the beach. We may be short of sunshine, but we're no different than anyone else when our lawns begin to brown and our gardens wilt.

Towns and cities are always hotter than the country. Tall buildings block breezes. Concrete and asphalt store heat overnight. Trees and shrubs, which shade the ground and moisten the air, are scarcer.

Heat waves can kill. A 1936 episode caused the deaths of nearly five thousand Americans and 780 Canadians, mostly infants and seniors. Another four hundred Canadians died of related causes, including, ironically, drowning at the beach or cottage.

A heat wave is defined as three or more days and nights with temperatures of 32°C/90°F or hotter.

The humidex, a numerical index expressing the combined effect of relative humidity and temperature, was first widely used in Environment Canada forecasts during the torrid summer of 2001.

As everyone knows, a trip to the beach is the best antidote for torrid weather. The sea keeps things cool. For this reason, St. John's and Vancouver have never had a heat wave—though Victoria has. Both cities have a deep ocean on their doorsteps. Deep water, salt or fresh, never really warms up, even in mid-summer. So it cools the land around it.

Nonetheless, all the Atlantic provinces except P.E.I. have endured hot spells, some lasting five days or more. Until the 1980s, our worst such spell was in August 1944. Whether global warming will result in Ontario-style heat waves is anybody's guess.

Environment Canada has put out a humidex advisory for most of New Brunswick for two days running and there are several groups that are at high risk from the heat...[namely] the elderly, infants and young children.

—Wayne MacDonald, MD, New Brunswick's chief medical officer, quoted in *Truro Daily News*, August 10, 2001

Myth,

Lore

& Science

I sometimes think weather was more fun a few thousand years ago, when people had some control. Instead of listening to weathercasters and worrying about global warming, they hired some gods to look after it, with suitable input from themselves, and went on about their business. The Egyptians had the sun god, Ra; the Babylonians had Marduk; the Greeks, Zeus; the Scandinavians, Thor.

Once you put gods like that in charge, all sorts of stories can be invented to explain the weather. For instance, the Norse explained thunder as Thor's wagon rumbling across the floor of the sky, and lightning as sparks from his giant hammer. The ancient Chinese had another explanation for lightning: the goddess Tien Mu was flashing her magic mirrors. As for thunder, it was cooked up by the god Lei Kung. The Japanese believed in divine winds or *kamikaze*—hence "kamikaze pilots," as the suicide dive-bombers of World War Two were called—for such a wind had flattened Kublai Khan's invading fleet twice, in 1274 and 1281, thus saving the islands from the Mongol sword.

Closer to home, the ancient Mexicans of the Toltec empire explained wind as the work of a plumed serpent named Quetzalcoatal, god of the morning and evening star and inventor of writing and calendars. The Salish and other West Coast tribes believed thunder was the sound of a great bird beating its wings—hence "thunderbird." For them and other early peoples, thunderstorms were religious events.

Like I say, weather was more fun back then. It was full of meaning and mystery. If one asked an ancient Salish or Mi'Kmaq elder about thunderstorms, he'd beguile you with stories of gods and sky spirit creatures who talked with humans. Electricity and atmospheric pressure would never enter his mind, nor would he speak of convection currents, ions bearing positive and negative charges, occluded fronts, or the like. No, he would leave you feeling like a participant in weather, and not a helpless spectator. Ask a physicist or weatherman, however, and you'll get a scientific explanation that is doubtless more correct on one level, but arid by comparison, devoid of drama.

Not to suggest that ancient weather mythology was full of benign spirits and cosmic harmony. In fact it was riddled with superstition and gory with human sacrifice—the downside of putting gods of wood and stone and metal in charge. It seemed a fair bargain. Despite our scientific know-how, we sometimes feel even more helpless than they.

At least they could expect results. If they needed rain for a newly seeded field, they did a snake dance. If they needed more sun to ripen the corn, they sacrificed something or somebody. If they wanted a flood to destroy their enemies, they stuck pins in a voodoo doll. And if the weather gods didn't cooperate this month, they would next.

But not everyone could deal with the weather gods. Serious dickering was reserved for shamans and priests—the ancient weatherpeople. Elders brought their request to them, and the shamans acted as brokers with the spirit world. (Is that why we still say "The weatherman calls for snow," instead of "The weatherman predicts snow"?)

The shaman might enlist the whole community to perform some act of worship or contrition. The Hopi of Arizona, a farming people, danced rain dances to a mystic serpent. Thousands of kilometres away, Australian aborigines shared a similar ritual. The Aztecs of Central America practised human sacrifice, usually at the spring equinox, usually a young virgin. The worshippers of Baal, sun god of ancient Palestine, performed similar rites, thus kindling the wrath of the prophets of ancient Israel.

Thus the First Book of Kings recounts how Elijah called down—or predicted—a great drought to punish the Israelites and their King Ahab for worshipping Baal and the fertility goddess, Asherah. Right on schedule, the drought came, lasting three and a half years. Elijah summoned the nine hundred and fifty prophets of these deities to the top of Mount Carmel, where rival altars were set up. Elijah challenged the Baal priests to call down fire from heaven to consume their offering. When their entreaties and self-mutilations failed, he called on God to do the same for him, dousing the altar, wood, and dead ox with water three times. Then God's lightning fell from heaven, consuming not only the prophet's sacrificial ox, but the stone altar itself. Only after Elijah and his followers had slain the priests did rain fall on the parched land.

It must have been deeply satisfying to broker the weather like that. We may scoff, but such methods must have worked often enough to keep the faith alive. A different kind of faith was to blossom from Europe's scientific revolution in the seventeenth century.

The Scientific Approach

The 1600s ushered in a profound shift in weather thinking. Science began to debunk the old myths. Inevitably, this spoiled some of the magic. Knowing how things work has its charms, but it's not the same as making them work. Anyway, as the Middle Ages waned and the Renaissance dawned, brighter minds began to probe the secrets of nature.

It began in Europe with the Greek Aristotle (384-322 BC). He wrote a long essay entitled "Meteorologica" (whence *meteorology*, the science of weather), which explained all weather phenomena as interactions of fire, air, earth and water. It was not bad. Nothing like it appeared for seventeen centuries, not until the Italian scientist/artist/architect/engineer Leonardo da Vinci (1452-1519) came along. Among the many instruments Leonardo designed and created was one to measure air moisture or humidity—the hygrometer.

From then on discoveries came thick and fast. Another Italian, the math whiz and star expert Galileo (1564-1642), invented the thermometer; a pupil, Evangelista Torricelli (1608-47), came up with the barometer, which measures the changing weight of the air pressing over our heads. With this instrument the Frenchman Blaise Pascal (1623-62) discovered not only that air pressure declines as one goes up a hill, but that it varies from place to place at the same level. Thus we have the *pascal* as a measure of barometric pressure.

English scientist and mathematician Isaac Newton (1642-1727) forever changed how we look at rainbows. He demonstrated that they are in fact white light broken into its component hues. Placing a wedge of glass in the path of a narrow beam of sunlight in a darkened room, he projected the resulting spectrum on a white sheet. He showed that suspended moisture droplets act as both mirrors and prisms to reflect and refract the hidden colours of sunlight. No doubt thousands of people had noticed this effect in icicles, glass, and oil on water, but Newton had the privilege of explaining it. It was a brilliant discovery; but it stole some of the magic.

The Anglo-Irishman Robert Boyle (1627-91) noted that air takes up less space at lower temperatures and higher pressures, and reduced this phenomenon to a mathematical formula which we call Boyle's Law. Taking the notion farther, the Briton John Dalton (1766-1844) showed that air temperature dictates how much water a given volume of air can hold. Meanwhile, France's master chemist Antoine Lavoisier (1743-94) proved that "air" is really the inert gas nitrogen blended with lesser amounts

"Kill a spider, it will rain within 24 hours." (Not as silly as it sounds. Insects and their kin are more active during low pressure conditions.) As children we loved fine weather, so we never killed a daddy-long-legs, an insect predator, if we could help it.

of oxygen, carbon dioxide and several trace gases. In 1752 the American Ben Franklin proved, using a kite, that lightning and thunder are electrical phenomena.

More magic vanished; but new vistas opened.

Scientists were now like bloodhounds onto a fresh scent. In England, George Hadley (1686-1768) explained that the earth's rotation influenced its wind systems. Not only that, said the Frenchman Gustave-Gaspard Coriolis in 1835, they are deflected to the right in the northern hemisphere and to the left in the southern, along with the ocean currents they push. The phenomenon was named after him.

The more they learned about weather, the more feasible forecasting became. For it began to look as if, in the middle latitudes at least, large masses of air of roughly the same temperature and dampness moved round the earth from west to east. That meant that if, say, there was rain in Spain, it could be expected to fall on France a few days later, and later still, perhaps, on Russia.

The problem was that they had no method of communication to carry this information far enough and fast enough to be of much use. But the need was great—both in peace and war, but especially in war. In 1854 an Anglo-French armada was bom-

barding Russia's Crimean peninsula on the Black Sea when a sudden gale came out of nowhere and wrecked most of the ships. That year, Britain's Meteorological Service was born. Likewise, America's National Weather Service owes its origin to the loss of 3,000 ships and 530 lives to storms on the Great Lakes in 1868 and 1869.

On April 1, 1960 the US launched TIROS, the world's first polar-orbiting weather satellite. A year later, President John Kennedy invited various countries to take part in a global weather network. From this came World Weather Watch, involving some 150 nations. Forecasters can now look up inside tor-

nadoes and look down on the lazy spirals of hurricanes in the making. They have at their fingertips tools undreamed of by Vice-Admiral Robert FitzRoy (1805-65), first head of the British Meteorological Department.

From such small beginnings came today's complex, world-straddling weather forecasting apparatus. It takes advantage of every modern technology: telephone, radio, airplanes, rockets, radar, high-altitude balloons, satellites, electronic computers, and scores of other things which scientists use to measure, map, and communicate weather data. It explores every possible avenue, from global warming to sunspot activity to borings from ancient trees, seabeds, and glaciers. It even examined a controversial theory put forward by James Gleick in his 1987 book *Chaos*. Called the Butterfly Effect, it postulates that tiny events can trigger large results, that a butterfly stirring the air in Peking today could change the weather in New York a few days later. For the future, the sky's the limit.

What Causes Weather Anyway?

I often wake in the night and listen to the wind. At times it sounds like surf on a distant beach. At other times it's like a large animal breathing outside my window. On summer nights it's a sad lover sighing in the lilacs. In early fall it's a kitten playing with the fallen leaves. In late fall it's a grumpy bear knocking things over, shoving things around. And now and then, mostly between November and March, it's a rampaging bull who shakes our old wooden farmhouse to its foundations and buries the lane under snow.

When we think of weather, we think of many different things: rain and snow and hot sun and clouds. One ingredient that's seldom missing is wind. What is wind? A child aptly defined it as "...like air, only pushier." That's it: wind is air moving from place to place. And the force that impels it is the same one that pushes water out of a garden hose—a difference in pressure. In the case of a hose, the pressure comes from a pump that creates a springy cushion of air in a tank. In the case of the atmosphere, the force comes from the inclination of dense cool air to flow toward light warm air. The greater the difference the swifter the flow.

Ocean of Air

We live at the very bottom of an ocean of air, the earth's atmosphere. Unlike the watery ocean, which is agitated near the surface but fairly still below, the earth's atmosphere is most active near the bottom. That's where most weather is brewed.

Jesus rightly called the wind mysterious. While its comings and goings can be sensed, it has no more substance than a ghost. We see it caress the spring grass and tumble autumn leaves across the lawn. We hear it swish rain against our windows and rattle the eavestroughs with volleys of hail. We

The wind blows where it wills; You hear the sound of it, but You do not know where it comes from, or where it is going.

—Jesus of Nazareth, John 3:8 NEB.

The ancient Hebrews used the same word for wind and spirit. Likewise, our English word "spirit" derives from the Latin *spirare*, to breathe—hence respire, inspire, and conspire (i.e., to "breathe together"), transpiration, and spiracle (an insect's breathing pore).

shovel the drifts of snow it piles against our front door. We feel its push on a hilltop and taste its saltness at the seashore. But the thing itself eludes us.

What if we could actually see the wind? Veteran US weatherman Eric Sloane tried to imagine just that:

If different qualities of air had different colours, the sky would look like the writhing flow of coloured lights in those old-fashioned juke boxes, never quiet, always mixing.
—*Look at the Sky and Tell the Weather*

If the planet were heated entirely from inside like a wood stove, and if every part received the same amount of heat, temperature and atmospheric pressure would be similar all over the globe, and no wind would blow.

You may be thinking, well, the earth is like a hot stove inside, a very hot stove, as volcanoes attest. Even so, the surface is cool, sometimes freezing cold, unless the sun warms it. The sun does warm it—but not equally. If solar heating were everywhere equal we would still have a windless world. But the sun has its favourites. Dark areas such as asphalt, forests, and oceans soak up the most heat, while light areas such as grainfields and glaciers soak up the least. Intermediate in absorption are potato fields, rock barrens and the like. These natural heating differences help set the winds in motion.

Aircraft pilots taking off from Fredericton, Charlottetown, or Halifax see this in action every sunny summer day. Balloons of warm morning air rise over forested hills, generating white clouds in the afternoon. The rising warm air, like the air over a hot stove, sucks in cool air below to replace it. This movement is wind. The same thing happens globally, magnified by differential heating between the equator and the poles.

Role of Latitude & Tilt

The sun is the engine that drives all weather. And as anyone who has been to the tropics knows, solar heating is not uniform over the planet. It depends on latitude and the earth's 23° tilt from its plane of orbit. In the tropics the sun beats almost straight down, while at the poles it shines on a slant.

Straight down is hotter. There is less of the earth's insulating blanket to penetrate, less heat is lost in transit, and, like a sunbeam focused with a hand lens, the energy is spread over a smaller area. Slanting sunlight, on the other hand, has to pass through far more air and dust, it loses more heat in transit, and arrives more spread out. (Air itself does not heat significantly.)

That's the main reason why the tropics are hotter than the poles. Moreover, polar icecaps reflect heat back into space.

Surface heat absorption at the equator, in fact, is four to five times greater than it is in our latitudes. And because warm air rises, warm air is continually rising and flowing poleward, to be replaced by cooler air flowing in underneath from higher latitudes. It's a bit like hot air rising above a stove and being replaced by cool air drawn in along the floor. This circulation spawns the planet's major winds.

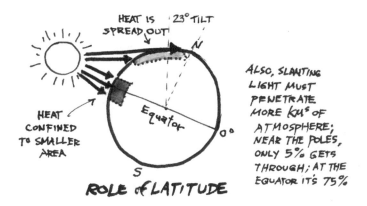

HEAT IS 23° TILT
SPREAD OUT

HEAT CONFINED TO SMALLER AREA

Equator

N

S

ROLE of LATITUDE

ALSO, SLANTING LIGHT MUST PENETRATE MORE KM^s OF ATMOSPHERE; NEAR THE POLES, ONLY 5% GETS THROUGH; AT THE EQUATOR IT'S 75%

Rivers of Air

I once climbed a 24m/80ft forest fire lookout tower with two of my children. At ground level there was barely enough wind to ruffle our hair. By the time we got halfway up, the wind was flattening our clothes to our bodies. Near the cupola, where the operator was shouting encouragement, it was hard to make ourselves heard over the wind's roar in the struts and stair rungs. At one point we nearly turned back. But once we reached the top, the view was magnificent. The descent was easy.

The kids and I had confirmed a basic weather truth—wind velocity increases with height. We don't notice this while going up in an elevator or a plane; but outdoors we do. Freed of friction with the earth's surface, air speeds up. Balloonists know this well. By ascending good and high, they make faster time. At 10km/6mi, they encounter the fastest of all earth's winds, the jet streams. Extending from that level up to about 12km/7.5mi, jet streams are flattish tubes of fast air hundreds of kilometres wide and thousands long but only a few kilometres deep. They form where cells of tropical and polar air meet in mid-latitudes. The

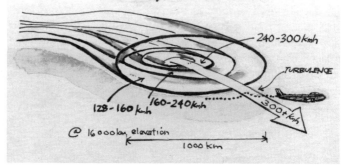

ANATOMY of a CANADIAN
(N. POLAR) JET STREAM

240-300 km/h

TURBULENCE

128-160 km/h   160-240 km/h

300+km/h

@ 16000 km, elevation

1000 km

three chief streams are a tropospheric westerly in our hemisphere, a tropospheric easterly in the southern hemisphere, and a mesospheric stream that switches from westerly in winter to easterly in summer. These three winds come in various velocities, being weakest near the equator and strongest toward the poles.

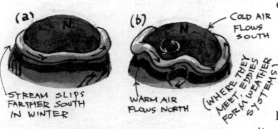

(a)

N

STREAM SLIPS FARTHER SOUTH IN WINTER

(b)

N

COLD AIR FLOWS SOUTH

WARM AIR FLOWS NORTH

(WHERE THEY MEET, EDDIES FORM WEATHER SYSTEMS)

NORTHERN HEMISPHERE'S JET STREAM: (a) STABLE PATTERN, (b) MEANDERING IN ROSSBY WAVES

The northerly jet stream that plays a key role in our weather blows constantly from west to east, looping farther south in winter. When it plunges abnormally far south, Florida gets snow and frost, citrus crops freeze and northerners pay more for their orange juice. At the core, jet stream winds travel at speeds up to 300kmh/180mph. East-bound commercial pilots tap into this, for it means faster flights for less fuel. But west-bound flights must choose lower levels or butt continual headwinds.

Highs & Lows & In-between
Atmospheric Pressure

Have you ever tried to remove one of those rubbery suction cups that hold knick-knacks to glass? It isn't easy. Often the only way is to lift the edge with a fingernail. The thing that holds them tight is air pressure. At sea level, the atmosphere presses on every surface with a weight of 1.1kg/cm^2 (15lbs/in^2). At this rate an ordinary paperback would weigh 150kg/330lb—if the air wasn't pushing equally in the opposite direction.

WIND SYSTEMS on a ROTATING EARTH

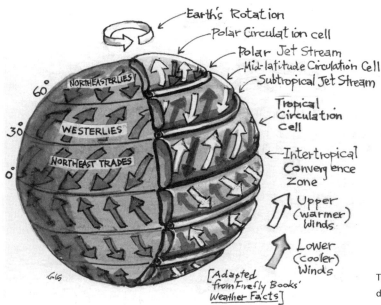

Earth's Rotation
Polar Circulation cell
Polar Jet Stream
Mid-latitude Circulation Cell
Subtropical Jet Stream
Tropical Circulation Cell
←Intertropical Convergence Zone
Upper (warmer) Winds
Lower (cooler) Winds

60°
NORTHEASTERLIES
30°
WESTERLIES
0°
NORTHEAST TRADES

[Adapted from Firefly Books' Weather Facts]

The phrase "in the doldrums" comes from the age of sail, when ships were often becalmed in the still air at the equatorial convergence zone, called The Doldrums. As Coleridge put it, "...as idle as a painted ship upon a painted ocean."

While air is light compared to water, it behaves in much the same way. Cold, dense air is heavier than warm, light air, and flows downward. Thus, when our bath water is too hot and we run cold water to make it comfortable, the cold water sinks because it's heavier. Likewise, on a frosty night it's the flowers at the bottom of a sloping garden that get hit hardest, while those at the top escape.

I've noticed this while cycling through hilly farmland after a hot day. The breeze feels warm on the hilltops and cool in the hollows. I've felt the same contrast while swimming in an ocean cove on a hot day as the tide turned. The surface water was warm, but underneath it grew suddenly cold.

Much like the currents and eddies in a river, domes of low pressure and valleys of high pressure move in a vast pinwheeling minuet from west to east. The lighter warm air ascends with counterclockwise inward winds (in our hemisphere), while the heavier cool air descends with clockwise outward winds, and each system takes about a week to pass any one point.

Because high pressure air usually means clear, dry weather, while low pressure foretells clouds and precipitation, meteorologists spend a lot of time measuring the two and comparing their fluctuations. And because range in pressure indicates the severity of the weather to come, they need to measure it accurately.

The Barometer

A math teacher once challenged his students to figure out the height of their ten-storey school building using any method they chose other than actually measuring it. One bright student got the correct answer by using an aneroid barometer. He took a reading at ground level and one on the roof, subtracted the upper from the lower, and used a weather manual to translate the difference into metres. The teacher had naturally hoped for a solution using geometry, but he gave the lad full marks.

Weather scientists calculate atmospheric pressure by the same basic method. The barometer (Greek *baros* = weight) comes in two basic types. The aneroid type uses a vacuum, and the mercury type uses a thin column of the silvery liquid.

At sea level a column of mercury typically rises to 1013.2 millibars (grams/second/centimetre), or 29.92 inches. The highest recorded air pressure (adjusted to sea level) was 1083.8 millibars (or 32 inches), recorded at Agata in Siberia on December 31, 1968.

The aneroid barometer (Greek *a* = no, *ner(os)* = wet), as its name implies, requires no liquid. Dating from 1843, it measures the effect of atmospheric pressure on a small, thin-walled metal box from which virtually all the air has been removed.

One side of the box is moveable and kept from collapsing by a strong spring, to which a pointer is attached. Rising pressure causes the side to buckle in; falling pressure lets it bulge out again. These movements are read off a scale in inches or millibars behind the pointer. By using a pen instead of a pointer, and having it press against a slowly revolving drum wrapped with graph paper, we get a record of pressure changes over time. This is called a barograph.

The mercury barometer was devised in 1644 by the Italian Torricelli. Taking a small container of mercury—a heavy metallic element which remains fluid at room temperature—he suspended in it a hollow glass tube with the upper end sealed. Torricelli wasn't surprised to see the mercury climb a few cen-

HOW AN ANEROID BAROMETER WORKS

HIGH AIR PRESSURE SQUEEZES BOX

LOW PRESSURE LETS BOX EXPAND

timetres up the inside of the tube, for capillary action does that to all liquids so confined. What did surprise him was that the column of mercury rose higher in fine weather, and sank lower in bad weather. By affixing a graduated scale he was able to read the heights from day to day.

When oldtime mariners said "The glass is falling," or "The glass is steady," this is the instrument they meant. It has saved countless lives on sea and land. Today Torricelli's barometer has been improved by adding a precise vernier scale and an adjustable mercury reservoir. Yet its principle is the same. It remains the single most important tool in the meteorologist's kit.

Still, a single barometric reading tells little. That's like consulting a compass after becoming lost. Learning where north is does little good unless you have *some* idea of your route. But if you've consulted the compass several times along the way, and if you know how to take a bearing and follow it, the instrument will lead you back out.

Likewise, weather forecasters read the barometer at set intervals. This tells them whether the weight or pressure of the overhead column of air is increasing or decreasing, and how fast. High pressure (denser) air flows toward areas of lower pressure like a ball rolling downhill. Just as the ball rolls faster the steeper the slope, so wind speed depends on the difference in pressure between two systems, called the *pressure gradient*.

By mapping readings from several stations for a given period and drawing lines between equal readings, meteorologists produce a chart of highs and lows that looks like the image to the right.

Because each line traces out a zone of equal pressure, it is called an *isobar* (Greek *iso* means "equal weight"). Adjusted to sea level and displayed on paper or on a monitor, it becomes a contour map of the air above the region at that hour. The closer the contours (the higher the pressure gradient) the faster the wind. That's why hurricanes, which are really rising whirlpools of warm moist air that suck in cooler air from around them, can pack such powerful winds. By plotting further readings at set times, weather experts create a series of snapshots of pressure changes at the bottom of the local atmospheric ocean. Such a chart, enhanced by readings for temperature, wind speed, precipitation and relative humidity, becomes a powerful forecasting tool.

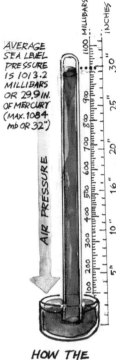

AVERAGE SEA LEVEL PRESSURE IS 1013.2 MILLIBARS OR 29.9 IN. OF MERCURY (MAX. 1084 mb OR 32")

AIR PRESSURE

HOW THE MERCURY BAROMETER WORKS

ISOBARS DEFINING HIGH & LOW

Role of the Earth's Rotation

If the planet did not turn, the wind would blow directly "down-hill" from high to low pressure. The currents would stabilize into a pattern of circulating cells looking something like the image to the left.

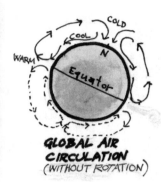

GLOBAL AIR CIRCULATION
(WITHOUT ROTATION)

But the planet does turn. You'd hardly know it, it's so big. Size and speed are relative things. An ocean liner moving 40kmh/25mph looks slow, while a speedboat travelling at the same speed looks fast. In fact a person standing at the equator, where the planet measures roughly 40,000km/24,000mi in girth, is travelling at about 1,670kmh/1,000mph. Even halfway to the North Pole where Maritimers live, the speed is about 1,200kmh/750mph.

To see the earth rotate, observe the sun near the evening horizon while holding your head still and reciting, "The sun doesn't move." You'll see the horizon roll up and cover the sun. Sunset has become earthrise.

Because the planet does move, the wind slants to the right in the northern hemisphere as the earth turns from west to east. In other words, the ball appears to roll around the hill, parallel to the isobars. This effect is due to the Coriolis Force.

Near the equator the fast-turning earth outpaces the air, creating easterly "headwinds" the way a cyclist does. In our latitudes the atmosphere's momentum "outruns" the earth, like a sled overtaking a child pulling it downhill. The resulting winds flow from west to east.

The Coriolis Effect

In other words, winds flowing up the sides of real highs or down the sides of real lows, often hundreds of kilometres in diameter, appear to spiral to the right. It's as if a person walked straight across a left-turning circular stage with paint on their feet. Their tracks would curve to the right, or clockwise. If the stage was a dome, the tracks would appear to spiral clockwise around the dome. But if the stage was a hollow dish, the tracks would appear to spiral to the left, or counter-clockwise. In the southern hemisphere these effects are reversed.

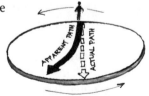

CORIOLIS EFFECT
(The Actor on the Rotating Stage)

The planet's rotation affects our atmosphere in another way. Centrifugal force—the force that makes a skidding car go off the road—pulls the atmosphere outward at the equator. Discus throwers use this phenomenon to add distance to a throw, and skaters use it to add speed to a spin. That equatorial pull gives the lower atmosphere a tiny midriff bulge about 16km/10mi thick, compared to half that over the poles. This explains why tropical clouds tower so high (they have more room), and why polar clouds are flatter and tend to hug the surface.

Another side effect of the earth's rotation is day-night heating and cooling. At sunrise after a clear night the ground is cool, having lost most of its heat into space. The slanting early sunlight is too weak and spread out to warm it much. By noon, however, the ground is noticeably warmer, and by midafternoon it's as warm as the season permits. Then it gradually loses its heat. The same thing happens to oceans, but they warm up and cool down far more slowly, thus buffering extremes of climate. Cloud cover slows the process, making the heat last longer. Without its daily solar sauna, however, the earth would steadily cool.

Surface heating also varies with the *albedo*, or reflectivity of the earth's surface. Dark surfaces (e.g. water, basaltic rock) absorb more heat than do lighter areas (e.g. wheat fields or glaciers). Snow absorbs hardly any heat, reflecting nearly all back into space. That's why a snowfield feels so warm on a windless day, allowing people to cavort in swimsuits. Oceans and lakes, on the other hand, absorb a great deal of heat. In fact, water has a remarkable ability to retain heat. And since moist air is paradoxically lighter than dry air, the world's water impacts powerfully on both temperature and air pressure, especially on the East Coast.

Clouds also reflect heat. When a cloud blocks the sun, we feel the chill. Basking in the first warmth of spring one moment, we're suddenly plunged back into winter the next. The lost heat is being reflected back into space, where there's no one to enjoy it.

Topography and elevation also play a part in local weather. A mountain's sunny side is warmer than its shady side, and its base is generally warmer than its peak. This sets up contrary winds. The rougher the terrain, the stronger the winds will be.

One summer afternoon on a forest survey in western Newfoundland, our group climbed the south face of 610m/2,000ft Killdevil Ridge in Bonne Bay. The temperature on the beach at its foot was a summery 21°C/70°F. During the two hours it took us to navigate the jumbled boulders and ascend, the air temperature fell steadily, until near the top we had to don our jackets. On the windblown summit we dug our standard soil pit. Half a metre down, we encountered a white clay that looked like ice cream and felt as cold. Its temperature turned out to be just above freezing—virtual permafrost at 49° N.

Because of the loose rock, Killdevil had no vegetation higher up. But across the bay on the 535m/1,760ft volcanic plug called Picket-on-a-Reef, sugary snow still lingered under the black spruce thickets; and nearby Table Mountain still sheltered old snow in north-facing ravines where sunshine never penetrates.

The Reason for Seasons

As if rotation and unequal heating weren't enough, there's the complication of seasons. "Thou hangest the earth on nothing," exclaimed the biblical Job nearly four thousand years ago. If the earth did indeed hang "level" in space like a Christmas ornament, we would have no seasons. To experience anything like one, people would have to travel north or south. The tropics would roast, our temperate zones would have endless summer, and the polar regions would stay frozen year round. Night and day would be virtually equal everywhere but at the poles, where perpetual twilight would prevail.

But the planet isn't "level." Like a spinning top that won't stay upright, it tilts 23° from the vertical as measured against its path around the sun. Thanks to that tilt, we get a bit of everything during its annual trip.

For example, in Nova Scotia, our day length goes from approximately twelve hours on March 21 (the spring or vernal equinox, which means "equal-night") to sixteen hours on June 21, the longest day. By September 21 (the autumn equinox) it is back to twelve hours, becoming eight on December 21, the shortest day.

As the tilted earth travels around the sun, Maritimers see the

The REASON for SEASONS

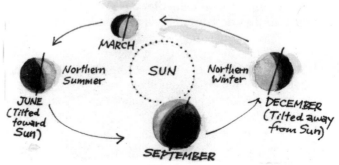

sun's angle at noon shift from 60° above the horizon in mid-summer to 30° in midwinter. At the spring and fall equinoxes it is 45°—which happens to be our latitude above the equator. In fact, sailors out of sight of land have for centuries determined their latitude by "shooting" the sun or North Star with an astrolabe ("star-take") or sextant and correcting for the date by using special tables.

On land, ancient peoples were aware that the sun rose and set in a slightly different place each morning and evening. In Nova Scotia the spring sun appears progressively farther north on the horizon, until it rises in the NE and sets in the NW. Then it appears to halt for a day or two. Astronomers call this the solstice or "standing still." After June 21 the sun begins almost imperceptively to "retreat" southward along the horizon. By late December it is rising in the SE and setting in the SW.

This apparent decline greatly alarmed ancient northerners. Druids and others offered sacrifices to bring it back. On December 21 or 22 it again stands still, then creeps north again. The date of Christmas coincides with an ancient festival celebrating this solstice.

By planting horizon stones to mark the middle (true south) and extremes of the sun's arc, and installing a central circle of sighting stones, the ancients devised calendars to regulate worship and communal life. Some, like Stonehenge in southern Britain, were sophisticated enough to predict solar and lunar eclipses.

Put it all together then—our tilted earth wrapped in its shimmering film of air and water, turning every 24 hours as it circles the sun at rocket speed once a year, its land and water surfaces unequally heated and swept by ceaseless winds, and we have the makings of all weather.

Now we need to look more closely at our ocean of air.

6

The

Atmosphere:

Womb of

Weather

God spreads the canopy
of the sky over chaos
and suspends earth
in the void.
He keeps the waters penned
in dense cloud-masses,
and the clouds do not burst
open under their weight.

—The Book of Job, Chapter 26:7, 8; NEB.

This Vital Vapour

Have you ever wanted to go to the moon? There's a good chance that this may be possible within a decade. I'm told the scenery will be astounding. No green vegetation or blue sky, mind you, just a black-white-grey landscape under a black sky peppered with stars, under a blazing sun with the blue-green earth rising and setting once every 24 hours. Weightlessness and walking in low gravity will astound you, too.

But the visit will have to be brief. The moon has no air. Unless the ship or its terminal can generate enough on the spot, it will be strictly rationed. A shuttle-load of passengers cavorting on the lunar surface would soon deplete the supply.

But then, airlessness sharpens scenery. The stars look brilliant against the velvety black sky. There are no clouds to hide them, no haze to blur the distant hills. On the other hand, even your insulated silver suit doesn't quite shut out the arctic chill when you enter a shadow, nor the furnace heat of full sunlight. Unlike on earth, there is no middle temperature.

You might notice other differences. If you waved your arms, there would be absolutely no resistance except the stiffness of the suit. If you lobbed a lunar rock into the lunar sky, it would fall without a sound, and the spurt of dust where it hit would settle instantly, as if magnetized.

Despite such novelties, before long the moon's sterility would get to you. You'd realize as never before how profoundly air contributes to life on our planet. Not just for breathing, but to shield us from the brutal cold of space, to buffer hot and cold, to carry music and human laughter, to float birds across the vault, to loft water vapour and dust into the sky for glorious sunsets. Above all, to colour the sky blue. By the time you were hurtling home, you'd be glad to see our friendly blue planet again.

For air is our native medium, the plasma of our lives. Unless it blows too hard or flings dust in our eyes, we don't notice it any more than fish notice the water they swim in. We take for granted the sounds it brings: rain pattering on a roof, birds singing at dawn, leaves rustling in the breeze, waves lapping a beach, ships' whistles echoing in the night. We take for granted the motion it imparts to things around us: trees swaying like ballerinas, sailboats leaning away from the wind, dandelion seeds

floating across a meadow, the manes and tails of galloping horses streaming behind them, quicksilver reflections on the lake,
surf charging up a beach. Nor do we think much about the making of sand dunes, or the sculpted arabesques of drifted snow, or,
for that matter, the origin of snow itself. Snowflakes are artifacts
of aerial updrafts.

But that's all academic. Without this vital vapour, nothing but
primitive bacteria and a few deep-sea tube animals could live on
earth. Humans can go without food for weeks, we can forego
water for two or three days, but without air we die in minutes.
Divers and swimmers never forget this.

The Nature of Air

What exactly is this gas, this stuff of weather? Pure filtered air is three parts nitrogen and one part oxygen with a smidgen of inert argon (0.9%), plus a trace of carbon dioxide, water vapour, and ozone. Unfiltered air, as a beam of light in a darkened room clearly demonstrates, also carries a cargo of dust and other things. These include pollen and fungus spores, tiny seeds, insects, spiders on silken parachutes, soot, and various other pollutants from human activity and volcanoes. Apart from the human pollutants, this mixture has remained essentially unchanged ever since blue-green algae and other plants started to breathe in carbon dioxide and to breathe out oxygen.

The Anatomy of Earth's Atmosphere

Most diagrams of the planet's envelope of air greatly exaggerate its thickness. Even *USA Today*'s prestigious weather guide shows a vastly overblown atmosphere dwarfing the earth and extending far into space.

Useful and comforting as this distortion is, it is totally misleading. The trouble is that it's almost impossible to depict accurately, in the same picture, something as massive as the earth and something as wispy as the atmosphere.

The truth is, our film of air is exceedingly, even alarmingly, thin—barely 400km/250mi deep. If the earth were an apple, its atmosphere would be thinner than the apple's skin. Compared

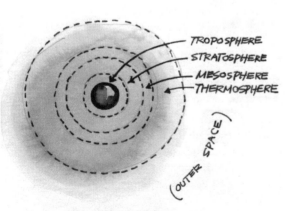

TROPOSPHERE
STRATOSPHERE
MESOSPHERE
THERMOSPHERE

(OUTER SPACE)

THE EARTH'S ATMOSPHERE:
STANDARD TEXTBOOK DISTORTION

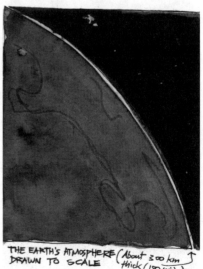

THE EARTH'S ATMOSPHERE
DRAWN TO SCALE (About 300 km
thick (180 mi))

to the planet's equatorial radius of roughly 6,400km/4,000mi, that's only 1/16. Worse, virtually all of the breathable air is in the bottom 5km/3mi. So the ratio is really only 3/4,000.

The reason why the film of useable air is near the ground and not diffused throughout the envelope is that air, light as it is, has weight and is compressible. It is subject to gravity. As gravity drags the molecules downward, they pack closer together. When forecasters refer to high or low pressure, what they really mean is the weight of the column of air above us. The greatest concentration of oxygen and the other components is at sea level.

If all the air in an average bedroom could be crammed into an average suitcase, a strong man could not lift it.

Just how much does the atmosphere weigh? As we have seen at sea level, it presses down with a weight of close to one kilogram per square centimetre or 15 pounds per square inch. That's a lot. If our bodies weren't already "pressurized" inside and out, we'd need pressure suits just like deep-sea divers.

We feel this on airplanes when a head cold makes our eardrums pop. During takeoff our throat-to-ear relief valve, the eustachian tube, gets plugged by normal mucus so that it can't equalize the pressure. Coming down is worse, because now the lighter cabin air filling our inner ear can't prevent the denser cabin air from pushing our eardrums in, which is painful.

This can happen even when high-flying airplanes maintain cabin pressure. The trouble is, cabin pressure is set to equal that at about 2,400m/8,000ft altitude, not sea level pressure. To match sea level pressure at high altitude might open cracks in the fuselage.

The need for pressurized living space is even more acute with space travel. Inside the ship, astronauts can work in ordinary clothes. Once outside, however, they must carry a sea-level environment with them, including air and warmth—hence the bulky suits. Stepping out suitless would not only mean instant freezing, but would cause burst eardrums and acute abdominal cramps as inner gases expanded.

The question arises: if the atmosphere is pressing down so heavily on everything, how come a jar of peanut butter isn't too heavy to lift? The answer is that the jar is being pushed from below and sideways by equal and opposing forces. But vacuum-pack that peanut butter, and its 10cm/4in lid will bulge inward under a load of 10kg/22lb. The jar itself won't weigh any more; the lid will just be harder to remove.

That *whoosh* we hear on opening a vacuum-sealed bottle of coffee is the sound of air rushing from higher pressure to lower pressure to equalize them. It is the sound of air going from one place to another—wind, in other words.

So here we are, inside earth's envelope of livable air, which does not even reach to the peaks of earth's tallest mountains, and which extends no more than 3m/10ft into the soil. A short distance above our heads—an average walker could cover it in less than two hours on the level—the air grows so thin, its oxygen molecules so scarce, that we would die there.

That's why alpine climbers at high altitudes must pause every few metres to "catch their breath," and why, when a plane's cabin pressure fails, passengers soon feel drowsy, disoriented, and racked by gas pains. It's also why airport runways must be longer at high altitudes, and why the long-winged condor, a kind of vulture, cannot fly above a certain height.

Up there, propeller aircraft with internal combustion engines falter because they need air both to float on and for the pro-

At 12,000m/39,000ft, air expands to seven times its sea-level volume.

pellers to bite into. Jet planes, which suck air in at the front and shoot it out behind, also perform best in dense air. And both systems need oxygen to burn their fuel. Giant balloons, which only require fuel to heat the air that gives them buoyancy, can ascend to roughly 30km/19mi. Space ships are able to leave the earth's atmosphere altogether—but only because they carry their own oxygen to burn the propellant.

Scientists call the bottommost layer the *troposphere*, and the zones above that the *stratosphere*, *mesosphere*, and *thermosphere*, in that order. Virtually all our weather takes place within the troposphere.

Balloonists have managed to reach the lower stratosphere, but above that balloons grow unmanageably huge and finally burst. The upper stratosphere is where space debris—sand, pebbles, and bits of nickel-iron,

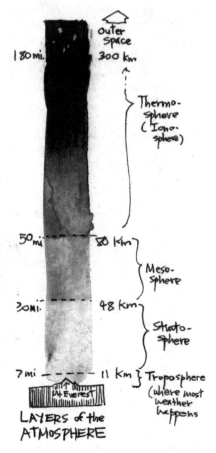

180 mi. · 300 km · Outer Space

Thermo-sphere (Iono-sphere)

50 mi · 80 km · Meso-sphere

30 mi · 48 km · Strato-sphere

7 mi · 11 km · Troposphere (where most weather happens)

Mt Everest

LAYERS of the ATMOSPHERE

what astronomers call meteors and everyone else calls "shooting stars"—burns up any clear night. The mesosphere, a transition zone, is bombarded by solar radiation. Here the aurora borealis (or, in the southern hemisphere, the aurora australis) performs its pastel extravaganzas, especially in winter. The thermosphere is so called because its boundary with the mesosphere is marked by a rise in temperature to about the freezing point.

In 1985 an American and French team finally located the sunken 1912 luxury liner *Titanic*. The broken ship lay in the cold and dark more than a mile beneath the North Atlantic. When they explored the wreck a year later with the camera robot "JJ", they did so from the safety of *Alvin*, a thick-walled submersible vessel equipped with lights and portholes. Outside, they saw bizarre marine creatures perfectly adapted to that hostile environment on or near the oozy bottom of the ocean.

I sometimes think of humans and other terrestrial life as deep-air creatures. Like the fish and crabs outside the *Alvin*'s portholes, we are adapted to life at the bottom of earth's ocean of air. Had the *Alvin* popped a porthole, those divers would have been crushed and drowned, like the crew of the Russian submarine *Kursk* in the year 2000. The pressure is too high for us, the oxygen supply too low, the place too chilly. To a fish, our world is just as hostile. The proverbial "fish out of water" suffocates in minutes. It drowns in air. A diver hauled too quickly to the surface suffers the bends, and a deep-sea fish literally explodes.

At 8km/5mi above the earth it is always winter. Far better, then, to live down here at the bottom of the ocean of air, blanketed from the chill of space, with all the oxygen we need, enjoying this amazing world of sound and motion and scent, and the blue dome over it all.

Ice Magic

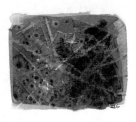

One cold grey November afternoon, driving south across western Nova Scotia from Annapolis Royal to Liverpool, thinking of ice and poetry and winter, I spotted a sign that said "Mickey Hill Interpretive Trail." This trail was new to me, so, needing to stretch my legs, I stopped to explore. Moments later I was crossing a bouldery chasm on a high, swaying rope bridge. From there the trail wandered downhill through tall old pines and oaks to a frozen cattail swamp.

Kneeling to photograph a lichen-clad boulder the size of a bus, I heard a rustle coming through the silent forest. White pellets the size of rice grains began to patter on the frost-crisped moss, the curled brown oak leaves, the glossy green polypody ferns. Confetti for the sad bride of winter.

First snow always gets me. It moves me more than March's first coltsfoot or September's first red leaf. For even as it hides the rubbish of summer, it speaks of another year written off, of earth's swift passage round the sun bearing us with it, of closure and an end to warmth and life—the big Full Stop.

On the whole, I prefer ice to snow. Ice is forthright, substantial, and dependable. You can drive a truck or a team of horses over it. You can saw it in squares and sell it to hot countries to cool their drinks. You can fish through it or make a windowpane for an igloo out of it. You can smash it on the ground for the hell of it (there's plenty more where it came from). You can play hockey on it or curl on it. You can melt it on the stove with a great fizzing of water and steam. Ice needs no shovelling; a sprinkle of sand or salt will do. Walking or driving over it is challenging, but skating or sliding over it is pure delight.

Certainly its comings and goings are no less lovely than those of snow. In crystal form it accounts for some of nature's most dazzling phenomena—sun dogs, moon rings, frozen waterfalls, icebergs, and, come to think of it, snow itself. And for those who don't get out much, it creates overnight masterpieces on the windowpanes of poor and rich alike.

They say the Inuit have more than a hundred names for what to southern Canadians is ordinary *snow*. Linguists and anthropologists make much of this. But I'm sure that Maritime motorists have nearly as many names, some unprintable here, for ordinary *ice*.

And no wonder. After enduring five or six months a year of what newscasters blithely term "occasional icy patches" or "intermittent freezing rain," after fishtailing and skidding times without number, narrowly missing this truck or that drop-off and sometimes not, what would one expect? It's a love-hate thing.

The other morning, scraping the knobby stuff from my car windows for the tenth time this month, I was moved to reflect on the difference between the ice on my car and the kind I knew as a child. The stuff I was struggling to dislodge seemed of a common grade, rough and unfriendly like the kind that coats the riggings of ships and capsizes them; the remembered ice was magical and somehow *amiable*. The actual stuff mocked my efforts and spoiled my morning; the imaginary stuff put a catch in my breath as I scrambled into my clothes at sunup on a Saturday morning.

Perhaps being wealthy helped—wealthy in ice, I mean. Most years our bay was covered with it for four or five months. This must have troubled our parents, for it cut them off from concourse with the outside world. To us it was sheer delight, another world to play in.

The two best ice seasons were when it was making and when it was melting. Most years it began to make in December. When frost ferns appeared on my bedroom window overnight, it was time.

On such a morning I'd head for the nearest ditch or puddle to check on progress. If the surface looked strong enough, I'd test it by jumping on it. If it didn't, a rock would tell how thick it was without my risking wet feet.

Something that always puzzled me was exactly how ice formed in the first place. Often at dusk in December I'd lie down by a puddle and try to see it actually *happen*. First there would be just the dark water shivering in the cold. Then, out of

Canadian snowboarders have coined some new words and phrases for snow:

Base - a fine hard-packed layer covering bare ground
Bullet-proof - Super-hard or frozen
Cement - heavy snow found in coastal regions
Crud - Varied, inconsistent snow

nowhere came these tapered little transparent points, feathered on the edges like tiny Christmas trees, creeping out over the quivery surface from all sides.

But they never *did* anything while I watched. They were either there or not there. Yet next morning the crystal matrix would be complete, its interstices all interlocked and cross-braced, a geodesic dome in two dimensions.

Very mysterious.

Just as mysteriously, after a few false starts, the whole bay would catch over from shore to shore, first in sheltered coves, then from headland to island, and finally down the center where the river current kept a line of dark water open. Until it closed we were forbidden to set foot on the bay proper. By early January most years, a team of work horses could safely haul a load of logs across. Then we didn't need permission any more.

8

Weather & Flora

Why Do We Have the Plants We Do?

The last place I expected to meet a yellow birch that August afternoon was near the windy summit of Picket-on-a-Reef, a cusp of ancient basalt jutting six hundred metres above Bonne Bay in western Newfoundland. Yet there it was, miles north of where it should be, anchored to an unstable talus slope at tree-line.

As a forester and naturalist, I've been struck time and again by how sensitive plants are to temperature and light. That dwarf yellow birch, growing at the same elevation in southern New Brunswick, or even at sea level here, would have been a sizeable tree. There was ample seepage of nutrients down the slope. There was ample sunlight. The problem was too much wind and cold. At that elevation and latitude, even hardy species like black spruce and balsam fir were barely waist-high.

Our part of Canada is in fact the northeastern rim of the world for many heartland plant species. Here their post-glacial expansion north and east, in progress now for over ten millennia, begins to falter. Butternut and silver maple, accustomed to softer air and longer days, drop out in southwestern New Brunswick. Black cherry, its blossoms repeatedly nipped by frost, has gotten only as far as western Nova Scotia. And while beech, sugar maple and eastern hemlock have managed to re-colonize all three Maritime provinces, they evidently arrived too late to cross the Cabot Strait into western Newfoundland. Red and white pine did so, likewise red maple and yellow birch; but black ash only secured a toehold in the southwest corner. The same applies to hundreds of kinds of smaller plants.

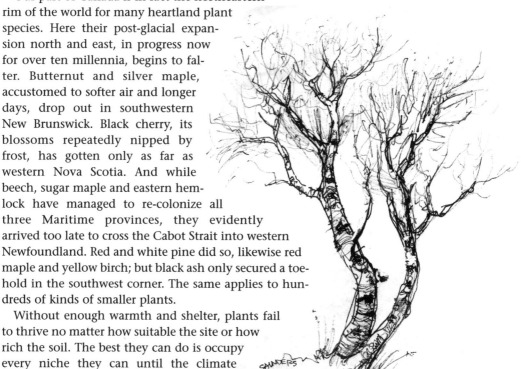

Without enough warmth and shelter, plants fail to thrive no matter how suitable the site or how rich the soil. The best they can do is occupy every niche they can until the climate changes.

ACADIAN ▥   BOREAL ▧

FOREST ZONES

The range maps in my tree books show this clearly. Every species has an optimum zone where it does best, and a peripheral zone where it does less well.

My runty yellow birch was really pushing its luck.

Any plant hardiness map shows this principle at work. Such maps are printed inside the covers of seed catalogs to help customers decide what to plant where.

These maps employ a matrix of risk factors to plot the likelihood of plant survival region by region. The original U.S. Department of Agriculture map had 11 zones. Canada has split each zone into (a) colder and (b) milder sub-zones. Within each zone, certain plants can thrive and others cannot. Thus a shrub okay for HZ 6 should do well in that zone, but not in a cooler or warmer zone. Horticulturalists and gardeners rely on detailed versions of such maps. Seed catalogue maps are so tiny that anything smaller than a county falls through the dots and pixels.

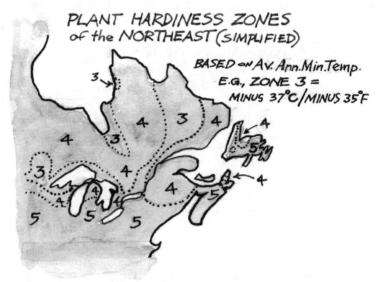

PLANT HARDINESS ZONES
of the NORTHEAST (SIMPLIFIED)

BASED on Av. Ann. Min. Temp.
E.G., ZONE 3 =
MINUS 37°C/MINUS 35°F

(Note: To access Canadian maps, try the worldwide web at http://res.agr.ca/cansis/systems/online_maps.html)

Still, they help in ordering plants from outside one's region, especially from nurseries and seed houses that rate their products by hardiness zone.

The plantscapes of the Maritimes have been developing ever since the last ice sheets melted. Throughout most of that time their dominant vegetation has been forest. Before that, our region was a barren rockyard of sand, mud, gravel, and boulders.

But time has draped our hills with lichens and mosses, clothed the slopes with forest, carpeted the valleys with grass and ferns, seeded the wetlands with cattail and cordgrass. The landscape has come alive, a symphony in chlorophyll.

Travel from the Gaspé to Cape Breton and you'll see a great diversity of forests, from upland sugar maple groves to black spruce bogs. Much of that diversity is caused by topography, nearness to salt water, soil type, and human impacts on the land. But the main thing is how cold the winters get. Other important factors are the length of the frost-free period, how much rain falls each summer, how much snow accumulates each winter, how hot the summers are, and how windy it gets.

In a word, weather.

I've travelled that route many times by train and automobile, both in winter and in summer. Always I've been struck by the shifts in vegetation, especially forest types. Because native trees have different needs and capabilities, they are reliable indicators of local climate.

The frost-free growing season around Yarmouth and in the Annapolis Valley is a month longer than northern Cape Breton's.

For example, along Bay Chaleur and in the Jacquet River country a sugar maple/hemlock/pine type dominates the hills. You really notice this in October when the maples are at their best. But inland, up in the headwaters of the Miramichi and Nepisiguit rivers, the hills have a northern look because balsam fir, pine, and birch thrive in the cool moist air.

Coming down onto the triangular rolling coastal plain that extends from Shippigan to Fredericton and from Prince Edward Island and to Pictou, we enter a landscape dominated by red spruce, hemlock, and pine, with occasional clumps of northern white-cedar. One finds a similar landscape in Nova Scotia's central and western interior.

Between Sackville, NB and Amherst, at the head of the Bay of Fundy, we see a ribbon of white spruce and balsam fir fronting both shores of this windy, tide-churned bay. Soaked by cold fogs and lashed by salty gales, they have a pruned and battered look. A similar forest borders Nova Scotia's whole southern coast, for the same reason. And behind each fringe of conifers rise rolling uplands with sugar maple and yellow birch on the slopes and fir on top. It's no coincidence that eastern Nova Scotia's lowland forest resembles that of New Brunswick's Restigouche country, nor that the Cape Breton upland forest looks like the Nepisiguit forest. Each has similar weather. Of course many of these woodlands have been logged and cleared and perhaps burned in the

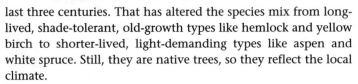

Sugar maple thrives on upland slopes mainly because there is less frost than higher up *or* lower down. Temperature falls with altitude, and cold air flows downslope like water, pooling at the bottom in "frost pockets." Fir thrives in cold and wet, so it forms the natural forest of Maritime highlands and frost pockets.

last three centuries. That has altered the species mix from long-lived, shade-tolerant, old-growth types like hemlock and yellow birch to shorter-lived, light-demanding types like aspen and white spruce. Still, they are native trees, so they reflect the local climate.

The Virtue of Staying Put

The great fact about plants is that they cannot move. Unlike animals (except for barnacles, mussels, and a few others that travel before settling down), they stay where they germinate. The only way they really travel is in the form of seeds or twigs, or by sending out roots and shoots.

Immobility means plants have to take what comes. What comes may be a gnawing mouse, a bud-snipping partridge, a lawn mower, a gardener with shears or hoe, a boy with matches, a man with a chain saw, an avalanche or landslide, a forest fire.

And weather. Day after day, night after night, year in, year out, they endure cold, heat, rain, sleet, snow, wind, thunder, lightning—whatever nature throws at them. They're out in all weathers and can never come in.

This was brought home to me during the summer of 1999. A friend and I were hiking around Long Point on Twillingate North Island in Newfoundland. Long Point is a finger of blue-black basalt pointing at Greenland with nothing but North Atlantic in between. In the lee of the lighthouse the June sun felt warm, but we were a hundred metres above sea level and a pack of icebergs chilled the air. Leaving the pavement, we took a path through a fir thicket to catch a view of Sleepy Cove boiling with white breakers. And there, out of the wind, half hidden among reindeer lichens on a smear of brown gravel in a bowl of rock, was a strange little plant. Its rosette of dark, incurled oily leaves hugged the ground for warmth. In the centre was a hair-like stalk bearing a single intensely-blue flower with a gold heart. The flower stirred in the fitful breeze like a little flag.

We'd both seen this plant somewhere before. I knew it wasn't blue-eyed grass, or a violet, or a very stunted blue-flag iris (of which there were many growing nearby). Later I looked it up in Roland's *Flora of Nova Scotia*. It turned out to be something much rarer, namely butterwort. Butterwort traps insects on its sticky leaves. Its Latin label, *Pinguicula*, is surely one of the loveliest names in all botany. This species is what botanists call an arctic-alpine species. It also grows in Labrador, on stormy St. Pauls Island off Cape Breton's northeastern tip, and in coastal New England. It was likely catching mosquitoes across the whole region when the glaciers were melting here millennia ago. Today's warmer climate has confined it to a few subarctic locations.

The dramatic effect of exposure was made plain to me one hot August afternoon in the Annapolis Valley of Nova Scotia. To escape the suffocating humid heat I drove up the 250m/820ft North Mountain and down to the Fundy shore. The trip took less than ten minutes, yet it felt like going from a furnace room to an air-conditioned restaurant. In five minutes I had to put on a sweater.

Like people, plants "feel" such changes and adapt accordingly. A few kilometres away on the valley floor, great elms and black locust trees overarched the streets, their leaves hanging limply in the heat. Here on the coast, stunted white spruce and balsam fir clung to the cliffs. I might have been on the coast of Maine or Newfoundland.

It also matters in what direction a hillside faces. A south-facing one is sunnier and warmer than a north-facing one that is in shadow much of the day, and the vegetation shows it. Other things being equal, you'll see more hemlock trees on the cool shaded north face of a hill than on the south side. The south side has what amounts to a longer growing season, warming up sooner in spring and cooling off later in autumn.

Among the toughest places for a tree to survive are offshore islands without protective headlands. One June in the 1970s, helping with a ptarmigan transplant to Scatarie Island off eastern Cape Breton, I was marooned when the wind changed to easterly and brought the drift ice back. With time to kill, I studied the flora.

Scatarie Island's eastern end is one windy spot. Trees have a hard time of it. The only specimens of any size grow in sheltered hollows or in the lee of cliffs. Balsam fir is the most common, and seldom gets taller than two or three metres. It must contend with not only a short growing season interrupted by icy fogs

December 13

And then a small overnight snowfall (8") preceded by freezing rain that bent down the softwoods. The branches of our Mugho pine, its blue-green foliage burdened with wet snow and enamelled with ice, sprawled in contorted brown loops almost on the ground. The white birch I planted 5 yrs ago lost its tallest shoot. Our grey 1988 Mustang encased in a translucent enamel that proved extremely difficult to dislodge. The ice would have been fully transparent but for a lovely surface patterning like that of morocco.

—Author's journal

and salty gales, but with a long winter of driving sleet and blizzards off the North Atlantic. As a result, the top of every tree is dead, its bark chafed down to raw wood.

Yet inside those tangled thickets there is hardly a breeze. Under the old trees' shelter I found normal young fir, perfect little Christmas trees. But I knew that the first winter after they poked their top branches up into the wind, every needle would be shaved off.

The point is, plants grow where the weather suits them. Which is another way of saying that, other things being equal, weather virtually dictates where plants grow. When Scotch broom showed up along Nova Scotia highways a few decades ago, people feared it would spread uncontrollably. But it seems that winter temperatures are confining it to the warmer western counties. The same applies to the province's native black cherry, and to butternut in southwestern New Brunswick.

Blowin' In the Wind: Seed Dispersal

Sit outside in the country some warm June evening and watch the dandelion seeds sail by. Lofted from ghostly grey plants with every puff of wind, each has a tiny parachute that carries them far. Some get tangled in trees, many drift out to sea—but enough alight on fields, golf courses, and lawns to ensure survival. Native plants like aster and goldenrod use the same strategy. Fireweed pods split lengthwise in August, curl four ways and release tiny seeds equipped with silken threads, which do the same job.

ELM (L.) MAPLE (C.) & ASH (R.)
SAMARAS

Many trees and shrubs rely on wind dispersal. Not apple, plum, cherry, and oak, of course—their fruits are too heavy. They must rely on birds and mammals—including humans—to carry their mail. Maple, ash, and linden have heavy seeds too, but they have evolved ingenious propellers that ride the breezes. For example, the samaras of maple act as helicoptor blades to spin the bean-like seed beyond the parent tree's shade. Ash trees use a paddle-shaped wing for the same reason. The linden, a common street ornamental, releases a strap-shaped wing from which dangle two or three berry-shaped seeds. White elm seeds have a wafer-shaped wing that floats on spring freshets. Poplars use both wind and water.

In fact, wind dispersal is fairly primitive. Fungi, ferns, and horsetails disperse their spores that way, and it is still the method of choice for conifers ("cone-bearers"), like pines and firs.

Tree cones consist of scores of triangular curved scales that spiral around a central spike and overlap like shingles. Each scale cradles a pair of triangular winged seeds. The cones, in full light near the tree's tip, ripen in early fall but remain closed until warm, dry weather coaxes them open. The seeds then tumble out and whirl away.

Dry weather also activates the explosive seed pod of *Impatiens* or touch-me-not. Despite the plant's common name, it can fire its seeds a fair distance without help. Witchhazel uses a similar system.

SPRUCE CONE

Using such tricks, plants spread their kind wherever they can find a suitable niche. In the twelve millennia since the last glaciers began to melt, they have reclaimed most of Canada. They do so by continually pushing against the limits set by weather and climate.

Highly evolved plants like burdocks and beggar's-ticks courier their seeds on mammal fur and human clothing.

Fall Colour

October visitors from Europe and Britain invariably *ooh* and *aah* over it. Back home, all they get are drab ochres and browns, and a scattered burst of pale yellow (e.g., Norway maple) or rusty orange (rowan). Nowhere else in the world except China and Japan do trees and shrubs put on such fireworks as do our maple, birch, and aspen. And while the show isn't purely Jack Frost's work as we once thought, it wouldn't happen without variations in weather. As the tilted earth "leans away" from the sun, less light and heat falls on each square metre of ground. It's a bit like an apartment cooling when the power fails.

The autumn colour shift is mostly a matter of subtraction. The leaf's green chlorophyll cells are withdrawn in response to dwindling light and heat, revealing its red, yellow, and blue pigments. As we learned in kindergarten, red and yellow make orange, and red and blue make purple. Purple hues are the rarest, being seen mostly in white ash and wild rhododendron, and in ornamental "copper" beech and maple.

By late September—sooner up north and at higher altitudes, later down south and lower—leaves begin to shut down. Like nervous stockholders, they withdraw all liquid assets (in this case, sugar-laden sap) and store them in a safer place, namely in the bark or underground. There the sap is infused with an antifreeze that limits the size of ice crystals which would otherwise damage the cell walls.

Finally, the tree seals off each leaf with a layer of watertight cork, and lets it fall. The fallen dry leaves in turn protect the roots from winter's freeze-thaw action. They also generate compost and provide habitat for small animals and invertebrates.

Meanwhile the hardwood hills and blueberry fields blaze with yellow (birch, aspen, elm), orange (maples, mountain ash) and deep crimson (maples, dogwood, wild raisin, oak). Most years the display lasts two or three weeks—again, depending on the weather. One strong October gale can strip a tree overnight.

Even without frost, broadleaved species still put on spectacular colour. Conversely, a dry fall dulls the effect.

Believe it or not, conifers also change colour and shed foliage in the fall. They're just less flamboyant. There are three reasons for this. First, only a fraction of the foliage is let go at a time, namely the older needles toward the trunk. Most are five or more years old. Second, the change is mostly hidden by the outer (younger) foliage. (Pines, with their open crowns and longer needles, take on a brownish look, as if the inner parts were dying.) Third, the colour is muted—bronze or dull yellow.

There is one native conifer which sheds all its needles in autumn, namely larch or tamarack (sometimes mistakenly called juniper). A school principal once asked me, during a discussion we were having with his grade six class, why the tamaracks were dying along the road to the school. He was a little embarrassed to be told in front of the class that this was perfectly normal for the species. Being a northern tree, it's usually the last to change, which makes it all the more conspicuous. Of our native conifers its green-gold foliage is the brightest of all.

The place to look for a conifer's autumn finery is *under* the tree. The ground beneath is sprinkled with gold. (Some still call conifer needles "sprinkles.") A pine plantation in late October can be ankle-deep in pink-gold which quickly fades to brown.

Spring Colour

Spring in the temperate forest is low-key if we hanker for bright colour. Still, after six months of winter drabness, it has beauties of its own. The colour is there, but more subtle.

Officially spring starts with the vernal equinox (Latin = *spring equal-night*) on March 21 or 22. At this time day and night each last twelve hours, after which daylight begins to gain on darkness. (The same thing happens in reverse each fall.) With increased light and warmth, trees and shrubs stir in their sleep. Roots waken in the thawing earth, sap rises through bark and branches, and leaf and flower buds start to swell. Weeks before, smart owners of sugar woods will have tapped their trees and readied their evaporators for another season of making maple

Flower buds

Leaf Buds

Winter Twig

After the buds burst...

syrup and sugar. By the time clouds of steam begin to billow from the sugar camp roof, spring is well under way.

As the previous year's buds swell and lengthen, far-off maple ridges blush pink, then change to jade green as leaves and catkins emerge. The subtlety of it reminds me of the colour play of northern lights. Where red or swamp maple is common, whole hillsides may redden with masses of tiny maple flowers—crimson for male, orange for the female, often on separate trees—followed by glossy beet-red juvenile leaves.

Conifers, being northerners by nature, wake up four to six weeks later than their broadleafed counterparts. As the upper buds swell and shed their onion-skin scales, out pop female conelets and catkin-like male flowers. From the latter come clouds of sulfur-yellow pollen, which the wind whisks away to fertilize other trees.

People sometimes ask, "What's that yellow film around the lakeshore this May? It looks like someone spilled paint or grease...." Closer inspection reveals a floating layer of pollen grains, each with a pair of stubby wings. Conifers—indeed all forest trees—put out bumper crops of seeds every few years, and in those years pollen is abundant.

Meanwhile, conifers sprout soft yellow-green shoots all over, briefly taking on a festive look. On fast-growing younger fir and spruce, notably Norway spruce, the shoots curve gracefully downward. Pines hold their new shoots upright in "candles." After a few weeks, the new needles form a waxy coating and darken to their normal moss-green hue.

To reduce self-pollination in species that bear both female and male flowers and cones on the same tree, the flowers or cones generally open a week or more apart. Also, in conifers the pollen-bearing organs are placed below the conelets. The immature cones of larch, spruce and fir, though too small to see from a distance, are a striking beet-red.

Pollination (Botanical Sex)

Through the spring and summer, an enormous amount of pollination goes on, both in the wild and in our gardens and fields. Except for a few self-pollinators like the common pea, all plants, unable to travel themselves, depend on wind or insects to fertilize their female flowers. Conifers rely on wind. But they are an ancient order whose ancestors flourished with the dinosaurs. Compared to the precision and efficiency of insect pollination, wind distribution seems primitive and wasteful. It takes so much pollen to do the job.

New tip growth

Older side growth with new shown paler

Perhaps that's why the vast majority of plants depend on airborne insects. To ensure that these couriers don't leave without a dusting of pollen, they have perfected elaborate mechanisms to attract exactly the kind of messenger they want.

Yet there is one very successful modern group which still relies on wind. Wheat, rye and corn all belong to the grass family, *Gramineae*, and all grasses are wind-pollinated. Without this partnership, you and I might have no bread or pizza or steak. Corn, or maize, is the world's chief livestock fodder.

One clue to this group's success may be that the grasses are natives of dry open prairie where wind is a continual presence and insect populations often get decimated by wildfire. Another is that grasses, like conifers, grow in natural monocultures. It's easier to deliver mail in the city than in the country because the houses are closer together. But insect-pollinated plants tend to be scattered among other species, so the sender needs a definite name and address.

I like to walk hedgerows in spring. Open to the sun and warm winds, they're the first to bloom. The same is true of south-facing roadsides, lakeshores, and the edges of cutovers and forest burns. Among the first are shadbush or serviceberry, wild cherry (three species), and trailing arbutus or mayflower. Close behind come red-berried elder, hawthorn, and hobblebush. All have white or cream-coloured blossoms. So do the later blueberried elder, mountain ash, sarsaparilla, and wild spiraea. The only early pinks are wild rhodora and laurel.

While day length triggers actual blooming, things like local temperature, cloud cover, and rainfall can hasten or retard the event. A plant that blooms in Yarmouth in April may wait till June in Labrador.

Thus, leaf-out and blossom-time are as much weather phenomena as spring showers. The engine that drives all this is of course the sun; but the sun's messenger is weather.

Seeing It Whole

I've often wished for a way to see the greening of the whole continent. The Canadian Centre for Remote Sensing in Ottawa has made that possible via colour satellite imagery. Those vertical pictures show the emergence of trillions of leaves on millions of trees, shrubs, and herbs from sea to sea to sea. Some of those images were published in the 70th anniversary issue of *Canadian Geographic Magazine* in January 2000. The play of greens and yellows and browns resembles a portrait being painted with pixels instead of brushstrokes. Seen in real time, it would resemble a green flush spreading from south to north and west to east. To my mind, these images rank with NASA's famous 1970s portrait of planet earth turning like a blue Christmas bauble in black space.

Through the weeks of summer and early fall, wildflowers wow us with other miracles of colour and texture. We've all seen the changing palette of June dandelions and July buttercups; of August's wild carrot, dog rose, violet, ox-eye daisy, touch-me-not, yarrow, lupine; of September's pink fireweed, yellow gold-enrod, and purple aster. All of them blossom and fade in step with nature's clock and with local weather.

Sit outdoors and watch, and nothing seems to be happening. The pace is too slow. But go away for a week and come back, and you'll be amazed at the changes—unless cold weather has inter-vened. The great New England naturalist Henry Thoreau kept meticulous records of the blossoming times of local plants around Concord, Massachusetts. After some years of this he made a boast. Set him down anywhere in the woods around Concord, Massachusetts without a calendar, he said, and he could deduce the date within two or three days. His confidence sprang from his habit of spending part of every day walking out-doors, observing flora, fauna, and weather. Most of us aren't so keen or so methodical. But it's nice to know nature is unfolding as it should.

Other Weather-Related Plant Phenomena

Cradle Hills On my first ramble in the woods of Nova Scotia and eastern New Brunswick, I was puzzled to see so many mounds and hollows. Except in rocky areas or marshes, they were everywhere. In slanting light they looked like waves on the ocean.

"Boulders under the moss," I thought. But they turned out to be made of earth and gravel. This puzzled me even more, until I found out why. They are the work of wind. The Maritimes lie in the path of fall hurricanes hatched in the Caribbean and the Gulf of Mexico. Although most hurricanes fizzle before reaching this far north, every few decades a big one destroys large tracts of old coniferous forest.

The mounds, locally called "cradle hills," are created when conifers blow over. Conifers have shallow roots that rarely—white pine is an exception—go deeper than 30cm/1ft. Moreover the root mass is compact, so that when a large spruce, fir or hemlock goes over, the whole mass is tilted clear and left up in the air—moss, soil, stones, and all. Next spring the mass thaws, lets go, and falls in a heap. This makes a nice dry island for seedlings to germinate on, since the hollows are often soggy from snowmelt. The resulting trees are even more prone to blowing down a few decades later. So the process repeats itself, and the mound rises higher. When the distance from hollow to crest is about one metre, the process stabilizes. This can take a century or two.

The Mechanics of Blowdown

In the summers of 1801 and 1802, with Napoleon Bonaparte rattling his sabre over Europe, Nova Scotia's Governor John Wentworth commissioned local naturalist Titus Smith Jr. to survey the interior of mainland Nova Scotia for naval stores. Specifically, his task was to find pines suitable for making ships' masts and turpentine for the Royal Navy, as well as lands suitable for growing hemp to make rope and caulking. In the course of his journeys, he saw vast tracts of tangled, blown-down trees. There were places where he and his guide could make no progress at all walking on the ground, but had to either crawl under the fallen trees or clamber over them. Sometimes it took them half a day to advance a kilometre or two. Some of the downed timber had been burned by lightning fires as well. They found little useable pine.

Then, as now, the worst blowdown occurred in coniferous stands along south-facing hillsides and shorelines where the trees had little protection from southwesterly gales. The same applies to old trees throughout the Maritimes, especially along the exposed Fundy, Northumberland, and Atlantic shores.

The worst damage occurs in spruce and fir, which have shallow roots, especially on swampy soils. Their dense foliage acts like a sail. Tamarack is less vulnerable because it sheds its needles in the fall (the only northern conifer to do so) and has a trunk that is flexible and tough. Pines are less prone to blowdown because their crowns are open and they favour well-drained sandy soils that allow deep rooting. And white pines, unlike most softwoods, develop a central taproot.

The term forest managers use for the toppling of trees by wind is *windthrow*. For example, "If we open that stand to the south, it'll be subject to windthrow." If windthrow takes place, the result is called *blowdown*, as in, "There's a lot of blowdown on Bob's woodlot."

One windy night in autumn, my woodsman father came exhausted to a trapper's deserted cabin and slept there. Sometime before dawn he woke, convinced the cabin was moving. He knew it had a reputation for being haunted, so he struck a match and looked around. He saw no ghosts, but the cabin was indeed moving. The floorboards rose and fell rhythmically. A gale had come up while he slept and, suspecting a connection, he took a flashlight and went outside. The cabin sat directly under a large white pine. He found that as the tree swayed in the gale, its stout roots rose and fell, lifting and lowering the cabin with it. Dad went back in and "slept like a baby rocked in a cradle."

In a typical windstorm a tree may suffer several hundred gusts an hour.

Being a tree in a high wind is like boxing with a powerful opponent with one's feet glued to the floor. The tree's only defence is to sway to and fro. Eventually, small roots snap and anchor roots tear away from the soil. Root rot and wet soil hasten this process, especially in old trees. Finally, overtaxed stem fibres start to snap like individual strands in a rope. The knock-out punch comes when a terrific gust strikes the tree before it can recover from the previous one. With a crash of breaking branches, the tree topples. If its crown is plastered with wet or frozen snow, or enamelled in ice, it will topple sooner, like a ship whose rigging is top-heavy with ice. Sometimes a falling tree drags several others down with it.

That's the most common scenario. The other is stem breakage, or *windsnap*. This usually happens when the root mass is frozen so hard that no flexing is possible down there. The trunk takes the whole brunt of the wind. Like a ship's mast, a tree trunk must be sound throughout in order to withstand such stresses. Even if it hasn't been weakened by heartrot, a point may come when it will snap. Normally, windsnap occurs within a few metres of the ground, where rot is likely to be more advanced and the trunk is stiffest. But if fungi have invaded the crown through a broken branch, the upper stem may go first.

In the first scenario, the tree nearly always dies. In the second, it may put out new growth below the broken top and survive for decades.

Most healthy coniferous stands weather big winds with minimal harm. They do this by growing in close-packed natural monocultures where each individual is of similar size and age, and can protect its neighbours. Trees that poke above the general canopy are also vulnerable, however. The older the trees, the more vulnerable they are.

But even young healthy trees can suffer damage if thinned too drastically, or if suddenly exposed by nearby clearcutting. That's why prudent forest managers prefer to cut even-aged conifers in stages. If they must clearcut—for example to harvest an old, vulnerable stand, to control an insect outbreak, or to salvage bug-killed trees—they try to avoid windprone areas and to leave buffer strips.

Even the most careful practices cannot guarantee against wind damage. One of the safest methods is shelterwood harvesting—taking out a third of so of the mature trees in two or three stages, thus protecting the remaining timber and ensuring natural regeneration of the same species. Even so, when we did a shelterwood cut on part of my woodlot several years ago, I worried about blowdown. Over the next few years we suffered some, but nothing serious, as we had taken the precaution of felling the aging and shorter-lived balsam fir beforehand.

But when native bark beetles began to show up a few years later, we had to speed the harvest in some areas, or risk losing many fine sawlogs to fungi that would invade through insect tunnels. Rather than clearcut, I opted for the seed-tree method. Essentially you clearcut in a good seed year, leaving six or eight windfirm, full-crowned trees per hectare. Ideally the seed trees stand long enough to restock the area, after which they are felled, or, if blown down, salvaged.

Sometimes the winds are so powerful that no tree is safe. This happened in New Brunswick on November 6, 1994, when sustained 100kmh/60mph winds from a large system walloped all of eastern Canada and downed an estimated 15 million trees in that province alone. In the area south of Mount Carlton Provincial Park, trees lay criss-crossed like toothpicks. Faced with this sudden unplanned harvest of three million cubic metres of healthy balsam fir and spruce—nearly half the province's allowable annual softwood cut—foresters had to abandon scheduled harvesting and direct all efforts to salvaging the wood. Nature doesn't let dead wood lie around very long. Wood-boring insects can smell dying trees from kilometres away, and the air is full of fungal spores in summer and autumn. Between them, they can reduce prime sawlogs to pulpwood grade within two years, and to worthless mush within five. (It makes good forest compost though.)

Deciduous trees are less prone to blowdown. They root more deeply than softwoods, and are leafless during the fall hurricane season. However, they fare less well during icestorms. This was obvious during the Great Canadian Ice Storm of 1998, when broadleafed trees suffered great destruction. This is because they are built differently than conifers. Conifers like spruce and fir, and to a lesser degree pine and hemlock, have a design advantage. They develop a strong central trunk supporting a cone of springy evergreen branches whose upper fronds overlap those below like shingles. In profile they resemble a very steep A-frame. Normally this allows the weaker upper branches to dump snow onto the larger lower branches until the mass cascades harmlessly to the ground. The disadvantage of this design is that the evergreen foliage provides a huge surface area for wet snow and freezing rain to cling to. Under such conditions the tree grows heavier and heavier, like a strong person bearing a tremendous load on their back. Even so, the trunk of a healthy, balanced conifer won't break. And if it is surrounded by coniferous neighbours, as spruce and fir tend to be, it won't fall. It's when it is exposed by cutting or pests, or weakened by rot or fire, that it topples. Windsnap usually occurs during cold winter with little snow.

The problem for deciduous trees is different. Since they are by definition naked in winter, their crowns hold little snow. Their disadvantage is that instead of developing a strong central column, they divide into several trunks. Species that tolerate shade, such as sugar maple, yellow birch, and beech, do so early in life. Intolerant or semi-tolerant species such as aspen, white birch and white ash do it later and so have a longer central trunk. Likewise, open-grown broadleafs are more branchy than forest-grown trees, and less protected by neighbouring trees. Unless the wood is very tough and bendable (as in elm and ash), or unless the branches have a drooping habit (as in ornamental birch), any overload of ice or snow tears them apart at the forks, their weakest point. Aspen, silver maple, and Manitoba maple are notably prone to winter breakage.

Though fall is our worst time for blowdown, it can happen in any month of the year. Summer thunderstorms can be especially deadly because they sometimes develop high-speed vertical winds that can descend without warning. Called *microbursts* (less then four square kilometres and lasting less than ten minutes) and *downbursts* (up to four hundred square kilometres and lasting more than one hour), they may pack winds of 250kmh/150mph on impact.

Plant life of some sort—grass, forage crops, forest, brush, heathland, bog—covers nearly every square metre of the northeast. And while the complexion of these plantscapes owes much to latitude, geology, topography, elevation, exposure, aspect, and soils, the dominant influence is weather. It decides whether a tree looks windblown or straight, at what time flowers bloom and wither, whether orchards will set and ripen fruit, when broadleaf trees will shed their leaves and how brilliant the fall colour will be—even the price of local plums in the supermarket.

9

Weather

& Wildlife

Stepping outdoors to do my stretches early one February morning, I caught a whiff of distant skunk, the first since the previous autumn. Not pungent as if under the patio, but faint as if from a field or two away. A not unpleasant smell, really—brisk and musky like certain perfumes, pungent like fallen leaves in autumn. But it caught me by surprise, coming so early, and because yesterday had been so cold. But overnight a warm front had slid in over the Arctic air, and the skunk, snoozing in her dark den somewhere nearby, must have felt the change. She stretched, yawned, and ambled off into the night to look for a drink of water and something to eat and maybe a mate.

Unlike that famous snoozer the groundhog, our skunks never really hibernate. Midwinter often sees them out in their striped pyjamas, like an insomniac raiding the fridge at two in the morning.

So it isn't strange that many people link the smell of skunk with spring. I know I do. Even without the hum of insects and the tinkle of birdsong, it gave me that feeling. But only for a while; a day later the temperature plunged, sending the skunk back to bed for a few more weeks. I kept rising by the clock and the sun; she waited on the thermometer.

Real spring is a five-sense experience and not really believable without a resurrection of animal life. Oh, the calico markings of old snow on tawny fields may delight us for a time, the yeasty smell of wet mud is thrilling, but one lone skunk isn't enough. Neither is a crocus or two in bloom, nor a few pussy willows, nor even a patch of green grass. What we crave is a toad to make us blink in surprise, a sudden glimpse of garter snake to hike our pulse a notch, a trout dimpling a stream, a migrant loon yodelling from the sky, a groundhog sitting still as a lawn ornament, munching new clover.

To appreciate the role of weather in the life of our wild creatures, from lowly worm to lofty eagle, we need only survey two seasons, namely winter and spring. How powerful that role is easily escapes us humans because we, of all nature's creatures, have largely escaped it. However, weather dictates to the wild animal, telling it where and when to travel, what coat it shall wear and when to shed it, what to eat and where, when to sleep and for how long, when they can mate and when they have young, and, finally, who will survive.

Found a woolly bear caterpillar near the shore, the larva of a tiger moth, a sign of autumn. This one wears the standard three-striped costume, black at either end with a rust-brown belt. The belt was fairly narrow, which is supposed to mean a cold winter...

—Author's journal, August 26

Lowly Worms & Others

One of the first signs of spring, real spring, is the reappearance of the creatures that scientists call invertebrates, so called not because they (the creatures) are inverted, but because they lack vertebrae. Wearing their "bones" on the outside, these mostly small critters inhabit the soil under our feet, clamber over the plants and buildings around us, paddle in the puddles and ponds, zip through the air—and provide lobster dinners.

For example, consider springtime's first outdoor spider spinning her first web to catch the year's first flying insect. Or the first moth struggling like a mummy out of its winter case. Or the first ant labouring up hill and down toting a pellet of food or an egg as big as itself. All these are animals. They are even (to use the latest definition), wildlife—i.e., life-that-is-wild. Invertebrates also include the spider-like mites and ticks, the many-legged centipedes and millipedes, the soft-bodied slugs, snails, and mud-worms.

Earthworms

Of all invertebrate wildlife, perhaps no creature so embodies spring as the common earthworm. Primitive as they are, without eyes or ears, last autumn they sensed the slow advance of lethal cold. Being small and slow-moving, they couldn't migrate south, so they migrated straight down. Earthworms "know" that the ground gets warmer downward, that below a certain depth frost never penetrates. Human miners know this too; the temperature several hundred metres down is tropical, year in, year out. (Farther north, of course, the ground freezes deeper; in the far north it stays frozen all summer, a condition called *permafrost*.)

Occasionally a northeastern winter is so mild that the ground never freezes, especially when snow comes early and stays. Snow being an excellent insulator, it helps retain the earth's "body heat." But this can't be relied on. So builders in frost-prone regions start every permanent structure by digging below frost level and pouring concrete footings. Without this precaution the building would tilt as the soil froze (expanded) and thawed (shrank). This would crack the foundation and cause doors and windows to stick; in time it might even topple the walls. We bury water and sewer mains for the same reason.

Millipedes, centipedes, snails, and many other creatures escape the cold in the same way. A notable exception is the monarch butterfly, which migrates thousands of kilometres to Mexico and back.

The temperature underground increases 9°C/48°F for each thousand feet downward.

November 17

Frosty dawn. Crows cawing overhead at sunrise, but no other bird sounds. In the pear tree, a mass of backlit (or one wouldn't see them) silken streamers, lifting and twining in the breeze. So that's what "light as gossamer" means. An unseasonal hatch of baby spiders, no doubt encouraged by the unusually warm air, a late coming. Invisible up close, but indubitably there, each thread has been loosed by a tiny spider programmed to launch downwind into life and either just landed or just about to leave. Will they die when killing frost returns?

—Author's Journal

A few insect species weather winter without migrating or burrowing. Ladybird beetles clump together by the dozens in some sheltered spot. Social insects like bees, wasps, and ants perish en masse, entrusting their future to a queen who is expected to survive and lay eggs next spring. Aquatic insects like dragonflies and caddis flies burrow into the bottom mud of a pond or stream. Luckily for them, water freezes from the top down.

Most other insects wait out the winter in an altered state. Two good examples are the cocoon and chrysalis, the solution of the butterfly and moth, respectively. Normally woolly or leathery, these mummy cases are usually so well camouflaged that predators like mice, chickadees and jays overlook them. Moth pupae often look like a rolled dead leaf or twig, complete with fake veins and colour dots.

Of course, once the caterpillars hatch into flying creatures they must contend with far greater dangers. At least now they aren't helpless. They can fly away, feign attack, look poisonous, or be so hairy as to be unpalatable. Sometimes they protect themselves with sheer numbers. Migrating moths of the spruce budworm, a voracious eater of softwood needles, sometimes blot out the full moon in July. When they settle to lay eggs, they may blanket many kilometres of forest with what looks like brownish snowflakes.

Seasonal migrations happen offshore too, though far less is known about them. Lobsters, a kind of giant, ocean-going insect, are a good example. As shallow coastal waters start to ice over in late autumn, they move to deeper, warmer water. Divers have met long queues of these crustaceans marching ant-like along the bottom, each sometimes (they say) holding onto the tail of the one ahead. Their destination is surmised to be some ravine near the edge of the continental shelf.

One such migration route seems to run from the Gulf of St. Lawrence through Canso Strait, the 0.8km/0.5mi channel separating Cape Breton Island from mainland Nova Scotia. For when the Canso Causeway was completed in 1955, the lobster fishery

The icy Labrador Current, flowing down a deep undersea channel between Labrador and Baffin Island, is rich in dissolved mineral nutrients and supports a great variety of marine life, including seals.

Insects that make sounds become mute below approximately 4°C/40°F and above 43°C/109°F. Grasshoppers can't fly in air below 7°C/46°F. Tame honeybees, while normally docile, grow irritable below about 11°C/51°F—perhaps because they can no longer defend the hive. Ants stay home when it's below 12°C/54°F. Above 40°C/103°F, they idle.

in Chedabucto Bay to the south collapsed and never recovered. Although the builders left a ship channel along the Cape Breton shore, apparently it was too shallow, cold or exposed for the lobsters to use. Unless they established another route out around Cape North, or found a substitute wintering place in the gulf itself, many of them may have perished.

Summer weather affects invertebrates in countless ways. The most basic is that, being "cold-blooded"—dependent on exterior temperatures—they can function only within a narrow range of cold and warmth. When it's too cold, they can't move. When it's too hot, their primitive tubal cooling system, encased in a suit of dark armour like a medieval knight's, can't cope and they must seek shade.

As temperatures rise, black field crickets speed up their territorial chirps. Next time you hear crickets in a movie night scene, count the number of calls in 14 seconds, add 40, and you'll have the scene's temperature in Fahrenheit degrees. (North American crickets don't do metric.) If it's a hot night, the chirps will be pretty fast.

July, for me, would be dull without fireflies. Not that I ever anticipate them. In fact, they always surprise me. Come a warm evening in July—seldom as early as June here—and they're back. After a long winter and what seemed like a longer spring, it was always startling, like suddenly seeing stars after weeks of cloudy weather, or the new crescent moon in the sunset sky. I'd be out for my night walk with the dog, jogging or walking along, thinking of nothing in particular. Out of the corner of one eye I'd see (or think I saw) one, two, maybe three winks of moving sulfur fire. At first I'd take it for a trick of light, a glint of distant yard light off my glasses. But when the roadside shrubbery came alive with intelligent sparks, now here, now there, now up, now down, there could be no doubt. The fireflies were back. Warming nights had roused them to court and spark as they have done in these latitudes since before there were humans to marvel at their cold fire.

Western Canadian artist William Kurelek captured a prairie version of the moment in his painting of fireflies darting through a dark cottonwood grove, with children in pursuit, a dance of summer.

In fact the firefly, though it flies, is no more a fly than firewater is water. It's a beetle of the well-named family *Lampyridae*, whose members possess luminescent organs on the belly. Species with wingless females and glowing larvae are called glow-worms.

Many people claim that mosquitoes, blackflies, and sand flies are more bothersome during damp or rainy weather. This makes sense, because warm, damp weather signifies low air pressure and thinner air, which makes it marginally harder for them to fly. It also signifies rain, which hampers feeding, so they're "stocking up for a rainy day." Also, damp hair tends to go limp and lie flat, slowing their progress over our bodies; and tiny water droplets on the skin cause them to cling more.

Conversely, cool high-pressure air makes flying easier. It also retards sweating and the release of body odours, which attract them in the first place. Low-pressure (warm) air increases both. And we know that body odours attract biting flies.

Spiders

As the days and nights cool off, spiders prepare for cold weather like everybody else. In August or September, numerous spiderwebs the size and shape of tea saucers appear overnight on lawns, parks, and golf courses. The sheet-web spiders that spin them hatch from the earth around the time of the first frost.

This allows the sheet-spinner to take advantage of the warm spell that usually follows—a time when most predatory birds have flown south and most insects have quit the scene—to have the last few flies to itself. These tend to fly low now, since the ground is warmer than the air.

Speaking of late autumn, one of its most magical weather-related phenomena is the "ballooning" or parachuting of spiders. Most people consider spiders dowdy stay-at-homes. Nothing could be further from the truth. Ballooning doesn't happen every fall, at least not in the same places; but when it does, it's fairly obvious. The perfect setting is a sunny afternoon with a warm breeze after the first frost.

On such an afternoon I saw, or thought I saw, what looked like strands of ultra-fine blond hair streaming from the lilac and spiraea bushes. The waving strands, back-lit by the sun, appeared and disappeared as the light played with them. I moved closer but still couldn't see them clearly. They seemed to be spiderless, so I assumed they were leftover lines from earlier flights.

One might assume this gossamer is spun by local spiders, but it isn't. It literally falls from the sky. Autumn is the time when many spiders hatch out those fat sacs which momma spiders lug around. Young spiders are programmed to release silk when they feel a breeze. A spiderlet on a rock or bush will pay out a length of web and then, like a bungee jumper, dangle in the breeze,

July 5

On my walk tonight, a half-hour before sunset, such goings-on.... Large water beetles, fighting or mating I don't know which; a male redwing blackbird whistling in an ash tree I raised and planted; wet mudflats shimmering, a dark brown insect with square wings that hid it and which from above looked very like a winged seed (an aphid?).... Alder bush lit by setting sun, the branches chestnut against dark violet and green background; on the saltmarsh. a young dead gull, its yellow beak hooked as fiercely as an eagle's, its feathers incredibly elegant, a study in pearl greys. On the way back, views of Old Barns in raking orange light. All in one evening, all lovely.

—Author's Journal

letting out more silk all the while. When this lighter-than-air parachute line gets long enough, the wind's tugging snaps the anchor line and the spider is off. If the wind is strong and a lot of spiders are trying to fly at the same time, the lines may tangle and fall to earth.

Ballooning spiders that reach the jet streams may travel for two weeks downwind before descending, logging perhaps a thousand kilometres or more. Ships 800km/500mi at sea have been showered with spiders. Airline pilots at high altitudes have reported seeing clouds of soaring spiders so dense they resembled snow. When the spider wants to descend, it rolls its silken lines into a ball; should it start to fall too fast, it simply releases more silk.

The proper name for these silken strands is gossamer, from the early English word "gos-somer," meaning "goose summer"—that warm interlude between the first warning frost and the later killing ones—the season we call Indian Summer.

Vertebrates
Amphibians & Reptiles
Next to earthworms and insects, the surest sign of real spring for most of us is the sound of spring peepers and other frogs chirping and trilling from recently thawed ponds and ditches where they have spent the winter asleep in the bottom mud.

Even before the frogs start up, the salamanders and skinks have responded to the temperature cue. Emerging some rainy night from their winter hideouts under forest logs and leaf mould, they waddle to the nearest pond or puddle to find a mate. Exact timing is crucial, or they risk freezing or starving.

Soon after the frog chorus starts, our several kinds of turtles and snakes, responding to the sun's caress, stir their cold bodies. Singly or in groups, they emerge from under old boards and rock piles, any safe and dry place. The snapping turtle, like the frogs and toads, hibernates in the bottom mud of the pond where it lives.

Fish
Where would you find a trout in winter? And would the trout take bait if you found one?

In autumn fish either migrate or slip into near-hibernation. An example of ocean migration was the mass movement in 1994 of capelin, a smelt-like fish, from off northeastern Newfoundland to the Scotia Shelf. Capelin normally migrate inshore to spawn on sandy beaches, but this move was unusual

Peeper rule of three:
Don't count on spring until
you've heard the peepers in
full voice for three
consecutive nights.

—James McConkey

and alarming. Biologists later discovered it was undertaken to escape abnormally cold water. The capelin returned north in 1996.

Of course, a fish living in a lake can't migrate very far. The most it can do is move from shallow to deep water or vice versa, seeking a comfort zone. And this is precisely what most freshwater fish do. Unlike frogs, salamanders and turtles, which can breathe through their skins, burrowing under the mud isn't an option. (In any case the mud would clog their delicate gills.) So they pass the winter under the ice in a state of near-torpor. While they do eat now and then, it's barely enough to keep their slow metabolism ticking over.

Astride Rennies River in St. John's, near the east end of Long Pond, stands The Fluvarium, a freshwater exhibit centre. A thick glass wall set against two metres of limpid moving water gives a unique perspective. What most intrigued me was looking overhead, seeing the water's surface from below, exactly as a fish might see it.

My visit being in late winter, the river here was still filmed with ice except in one or two places. From below, the snow-covered ice resembled greenish linoleum swirling with watered silk patterns, milky where thin, darker where thick. Through the open patches the sky shone like a quicksilver mirror. I nearly forgot to look at the trout. There were several of them, big German browns and smaller brook trout, hovering less than a metre above the pebbly bottom, where the current was less. Except for the rhythmic working of their pink gills and an occasional tail-flick to keep their place, their bodies were as motionless as dirigibles. They paid me no heed and seemed asleep.

There was no provision to feed them; in that state they wouldn't have needed much food. Had a worm or a beetle drifted down from their sky, no doubt one of them would have grabbed it. It was, after all, nearly spring. Ice anglers catch trout that way. But when heavy snow blanketed their roof, their sleep must have been almost true hibernation.

In choosing to escape winter this way, freshwater fish and amphibians take advantage of a most unusual trait of water. Like other liquids, water contracts when cooled. But then, at 4°C/40°F, just before reaching the solid state we call ice, it expands. No other substance except bismuth does so. Expanding, it becomes more buoyant. This causes it to rise. At the moment of forming ice, it expands further. That is why

There's a saying: "Wind in the west, fishing is best; wind from the east, fishing is least." Easterly winds usually bring rain, which drives insects to cover, so fish don't expect food.

Another old adage says, "The lower the barometer, the better the fishing." During a period of low pressure, if it isn't raining, insects will tend to fly low where the pressure is higher.

And: "A wind from the south blows the hook in its mouth." This refers to late fall, when the water has cooled. A south wind with sunshine warms the surface layer and may reactivate the fish—if you know where to look.

water, whether in a pail or a pond, freezes at the surface not at the bottom. A good thing too, since few living things besides bacteria could survive our winters if the reverse were true.

Come spring, the process is reversed. This is another good thing, for as the aerated surface water descends, it recharges the stale depths with oxygen to revitalize the bottom ecosystems.

Birds

Nothing demonstrates weather's power over wildlife more clearly than bird migration. The very phrase calls up images of Canada geese in noisy squadrons overhead. For people living along the eastern flyway, they are icons of seasonal change.

But geese are merely the most spectacular of avian commuters. Scores of other species, as large (like the common loon) or larger (like the great blue heron), wing the same airways with far less ballyhoo. These include robins and other thrushes, grosbeaks, many sorts of sparrows, various warblers and flycatchers, several blackbird species, the belted kingfisher, martins, swifts and swallows, and, amazingly, tiny mites of wrens and hummingbirds.

Some smaller birds jump the gun. Perhaps needing more pit stops, they tend to depart before summer ends. A good example is the semipalmated sandpiper. As early as mid-July, they arrive in the upper Bay of Fundy to refuel and rest. A farmer may be out haying, when suddenly the sky is full of tiny forms that wheel and bank as one, now flashing their white undersides, now showing their brown backs, settling along his windrows like autumn leaves as they wait for the tide to fall.

Fundy is about a fifth of the distance between their Hudson Bay breeding grounds and their South American destination of Surinam, about 5,000km/3,000mi away. It's a vital stop because Fundy's vast chocolate-brown tidal mud flats teem with newly hatched mud shrimp called amphipods. Ten days of gorging between high tides, and our long-billed bantam pipers have doubled their weight like prizefighters. Then one night they climb into the stars and wing non-stop to some tropical mudflat.

As the planet tilts toward our northern autumn, hordes of other birds pass like this while we sleep. They may alight briefly in tens or hundreds to feed and rest and twitter, but mostly we see them flying, high up where the winds are swifter and the friction less, where they can get their bearings from Polaris or the planet's electromagnetic field.

Just a bit more
and cold
might catch this
crow half wing
halfway across
slides
a glacier
left leaving.

—Peter Sanger from "Crow in Winter"

in *The America Reel*, 1983

Compared to the mass departures of autumn, the arrival of springtime migrants is a tentative thing: a few male robins here, a dozen sparrows there, two or three hawks sailing over, the cry of a loon too high to see the crier. Old Man Winter, instead of cracking the whip behind them, is retreating sulkily before them. So they dawdle. This makes sense, for food is scarcer now. Should a robin arrive too soon, it may freeze or starve—or both.

Yet, like a tree greening up leaf by leaf, it happens. One morning you step outside and the air is full of melody. Not the tiresome winter gossip of crows and jays, but something beautiful you haven't heard for months, something which the ocean of air delivered to your door. Spring without birdsong is unimaginable. It lights up the season as flowers light up summer.

Faced with travelling such long distances, why don't our summer birds stay all year? First, because there's not enough food and cover to support the residents *and* migrants all winter—no insects, few seeds, and far fewer succulent greens. Second, few migrants could survive the harsh weather. Even resident birds equipped with superb insulation (like eider ducks) and assured of high-protein food (like woodpeckers) have it rough. Third, birds can't hibernate. One price they pay for the ability to fly is a higher body temperature and a faster heartbeat than other creatures. Their metabolism can't throttle down that far. Also, if they store much fat they'll be too heavy to fly.

The ring-necked pheasant, an Oriental cousin of the domestic fowl (itself from southeast Asia), teeters on a climatic knife edge in the northeast. If the snow is too deep, or coated with ice for a week or two, it may starve. That's when they show up at patio feeders. By contrast, the native ruffed and spruce grouse, adapted as it is to sleeping in the snow and nibbling tree buds, fares very well. So do chickadees and nuthatches, which live mostly on high-octane foods like insects and their eggs, grubs, and pupae. Likewise crows, ravens, eagles, and jays do all right eating road kills, dead fish, nature's victims, and offal from slaughterhouses and farms. Owls and goshawks get most of their protein by killing grouse, hare, and mice.

Some birds change colour when snow comes. The snow bunting or snowbird, a brownish northern finch, goes from mostly brown to mostly white and back. Because it usually arrives here with the first blizzard, some nickname it "snowflake." The native rock and willow ptarmigan of Newfoundland and the northern mainland do so in lock-step with the failing sun and looming winter—first a few patches of white mixed with rust-brown, then calico, and finally pure

A patch of oil-soaked feathers the size of a quarter on an eider duck can kill it through hypothermia within a few days. It's like a snowmobiler with a hole in her jacket. The body heat leaks out.

Why do snowbirds always seem to arrive right on cue with the first big blizzard? The answer is, they were here already. Snowbirds summer in the Arctic and winter along our coasts. Storms drive them inland; afterward, they go back to the coast, which is milder and offers more food.

Swallows and bats, having extra-sensitive ears, are quick to detect a drop in air pressure. So are insects. All three then move closer to the ground or water, where the flying is easier.

white, even to the feathered toes. Its only dark spots are the beak and eyes and a flick of black on the tail.

For the majority of birds, migration is the only option when winter threatens. So weather controls their movements. That being so, birds have found ways to exploit the very forces that drive them north and south each year. One of these forces is air pressure. As bush pilots know, it's easier to take off and land in cool morning or evening air because such air is denser; a plane floats better in it. On the other hand, thin (warm) air is easier to fly through. Migrating geese fly closer to the ground in rainy weather, higher when it's cold and dry. Birds seek a balance between the high and low pressure.

Seagull, seagull, sit on the sand;
It's a sign of rain when you are at hand.

Barometric pressure falls before rain, making the air slightly less dense, less buoyant. This makes flying a little harder, in the same way that long-distance swimmers find fresh water a trifle more tiring than salt water. Besides, birds prefer to ride out storms on the ground. One way they get around this is to ride updrafts. Updrafts rise where the ground is warmer than its surroundings, or where a hill or other obstacle forces it to rise. Eagles and turkey vultures do it all the time, making "lazy circles in the sky."

The difference with migrating birds is that they try to ride only north- or south-tending currents. Where a mass of cool dense air collides with a mass of warm light air, a long ridge often forms along the line of contact, with updrafts on the warm air side, downdrafts on the other, and a steady wind along the crest.

When the ridge is aligned north and south, the birds coast along it. Since a north-south ridge means clear weather coming from the west, and east-west ridge means dirty weather from the east, this American saying about migrating geese is well put:

South or north, sally forth;
West or east, travel least.

If your chickens run for cover when it starts to rain, the shower will be brief. If they stay outside, it is going to rain all day.

—Lewis and Paulette Ramsey

Summer, despite its abundance of food and cover, affords birds new challenges and new opportunities. Rain soaks their feathers (unless they're waterfowl), sometimes chilling them if cold air follows. Mother birds try to protect their babies by nesting in dense foliage and by sitting over them during storms.

On the other hand, if the rain is warm, bathing becomes possible, a luxury most species indulge in whenever they safely can.

Mosquitoes and other biting flies harass them—though not as fiercely as they do mammals. In dry weather, birds can use dust-baths to kill lice and mites. They can also practice "anting"—placing live ants under their feathers to stimulate the skin.

Mammals

Unlike the lower animals, mammals must face the weather, or hibernate. All our native land mammals but two, namely caribou and bats, have opted to stay home. As for our sea-going mammals—seals, dolphins and whales—they are long-distance wanderers like the birds, moving north and south in spring and fall to breed and feed. The woodland caribou herds of Newfoundland and Labrador migrate scores of kilometres every spring and fall. The first snow flurries send them south to winter on windswept coastal barrens where snow is thin and lichens plentiful. Come spring they return north to graze in lush river valleys, pausing en route to calve.

As everyone knows, bears and groundhogs take a deep sleep that normally lasts all winter. That makes them true hibernators. Squirrels and porcupines aren't true hibernators, since they become active in mild spells. Skunks we've already met. Muskrats and beavers are a special case because they build winter lodges and stay active all winter without going outside much. These lodges are heated by the occupants' bodies and by latent heat from the surrounding water.

Mice are another special case. By the laws of physics any animal so small, i.e., with so high a ratio of surface area to body mass, should freeze to death. As a rule, the bigger the animal, the better it can stand the cold; most Ice Age creatures were big. Yet several members of the mouse tribe—notably voles—are active all winter, mostly under the snow, which is a good thing, because voles are the single most important food item for our predators.

How then do mice survive the cold? They tunnel under snowdrifts. Here they find not only warmth and light, but protection from above. However, coyotes and foxes can smell them through deep snow and will dig down to get them if they can. The perfect winter for a vole is deep snow followed by rain and a hard freeze. This renders the surface rock-hard and impregnable. As long as there is snow cover, wild mice go where they please in safety, digging new tunnels, nibbling grass and bark and seeds, mating and having young, generally having a good time. The only real menaces are the weasel and the shrew, both of which are small and fast enough to catch them.

American opossums live only as far north as their hairless tails and ears will permit. So far no wild 'possums are known to inhabit the Maritimes. Our winters freeze those delicate organs.

GLS

Hoofed animals, even tame ones, usually keep their tails to the breeze when grazing. This is so that if danger comes they will see it. Danger from behind doesn't worry them because they will catch a scent long before they would see an enemy anyway.

Checked the woodlot for
wildlife tracks in the
weekend snow. Plenty of
rabbit sign in the alder
section, but none under the
spruce—too exposed. Red
squirrels busy along the
edge.

Deer tracks everywhere,
especially in the ravine and
beside each fallen tree. Took
me some time to figure out
what they were doing. A
storm three weeks ago
broke the tops off several
old white spruce and blew
down some others. Could
see scrape marks along the
horizontal trunks and scraps
of lichen on the snow
among fallen needles, cone
scales, and dead twigs. The
deer were feeding on old-
man's-beard lichen.

—Author's journal

It is these mouse trails, trodden harder than the surrounding snow by countless hurrying paws, which show up as raised trackways branching over the ground like white roots after the snow melts. Until the snow returns, the mice must find other cover. Nothing is perfect.

Like the wild mice, our several species of moles and shrews also stay active all winter. But they tunnel in the earth, taking advantage of its protection and natural warmth to hunt worms and insects which are there for the same reason.

Larger plant-eaters like hare, deer, and moose cope with the hungry months by fattening on greens and roots in late fall, and by subsisting on twigs and buds, plus roots and greens from under the snow, all winter. In a pinch they will eat evergreen foliage. One adult deer eats roughly 36 litres (1 bushel) of live twigs every 24 hours, a moose several times that. If hunger strikes, they live on stored fat as long as it lasts.

While these mammals don't migrate, they move around. Deer and moose "yard up" (i.e., congregate in traditional areas offering ample browse with shelter nearby) during heavy snows. In late winter deer move to south-facing slopes where sunny days are warmer and snow melts sooner. Otters make regular circuits of favoured watersheds throughout the winter, thereby avoiding depletion of scarce food in any one area.

All this activity is hard on fur coats. Like the birds, mammals moult twice a year—a thin coat for summer, a thick one for winter. The family cat and dog retain the habit. With some wild mammals the shedding also brings a colour change that may be slight or drastic. White-tailed deer go from tan in summer to grey-brown in winter. Weasels and hares go from brown to pure white in stages. Day length triggers these transformations; the presence or absence of snow may play a part as well.

Of all our mammals, bats would seem to fare the worst, since insect food is very scarce from late fall to early spring. Since they alone of all mammals can fly, they migrate and hibernate both. Before the hard frosts hit, thousands of these mouse-sized creatures head for traditional *hibernacula*: abandoned mine shafts, dry wells, and especially caves. They seek such cold, dark, and damp places for three reasons. The cold chills them into dormancy, the darkness helps them sleep, and the dampness keeps them from drying out; they survive on stored body fat. Dangling head down by their hind feet, their velvety fur beaded with dew, bats slumber from October into April. Outside, blizzards and ice storms may be raging; inside, all is peaceful.

Weather
& Work

As a young man, I cruised timber for two summers on Newfoundland's easternmost extremity, the Avalon Peninsula. Avalon...a lovely Celtic word for the island paradise of the western sea whither Arthur and his knights were borne after death.... Well, it *is* in the western sea, and it *was* pleasant by times. It was also the wettest place I've ever worked in.

The peninsula is a lopsided star of ancient purple rock mantled in lush spruce and fir, and peppered with bogs and lakes. A wasp-waist isthmus joins it to the main island. Its two northern fjords, Trinity and Conception, face the cold Labrador Current, while its two southern bays, St. Marys and Placentia, open on the slightly warmer Gulf of St. Lawrence. With no part of it far from salt water, the Avalon is a playground for foggy rival winds, especially in spring and autumn.

Mare's tails make tall ships carry short sail.

Our job was to map the woodlands by species, age, and yield. Every fortnight we struck camp and set up a new base line from which parallel survey lines were run. Travelling by truck, canoe, and on foot, sleeping under canvas most nights, we had every reason to watch the weather.

The problem wasn't wet clothes. East Coast woods workers get wet anyway, even in fine weather, for cool cloudless nights produce the heaviest dews. In fact, dry bushes in the morning are almost a sure sign of rain before dark, while dew-soaked bushes betoken fine weather. No, the real problem was how to keep our maps, air photos, and tally sheets dry. We had acetate covers and clipboards, but soggy paper doesn't take pencil marks very well. A few hours of wet, and our notes were unreadable.

So the weather dictated our work schedule. Faced with so much rain and drizzle, we worked when it didn't rain, weekends included. Waking to golden sun on the tent meant two or three days of extra hard slogging through scratchy thickets and wet bogs, swatting moose flies and mosquitoes. Waking to the patter of rain wasn't so bad, for we could go back to sleep. The system worked as long as the weather cooperated.

However, sometimes the Labrador Current and the Gulf Stream brewed up a prolonged spell of wet. Keeping an old fashioned cotton canvas tent dry for day after day of close living was difficult. One touch from inside and even weather-treated canvas will weep.

One can only amuse oneself so long, reading, playing cribbage and guitar, and doing pushups. One's nerves begin to twang. Too much of a good thing becomes a bad thing.

One night I was jerked awake by what sounded like someone banging a piano with a saucepan. Lighting a soggy match, I saw my guitar, which I'd hung from our indoor clothesline to avoid getting it wet or broken, with its bridge unglued and dangling by six strings. I screwed it back on, but the guitar never sounded as sweet afterward.

Outdoor workers can't afford to ignore what the sky is doing. It affects their work and often their safety. That goes not only for farmers, market gardeners, seamen, and forest workers, but also for pilots, construction workers, house painters, roofers, and train engineers, to name a few.

This is something we share with our predecessors the Mi'kmaq, Maliseet, Innu, and others. Weather ruled First Nations people, whose work centred on food-gathering. It also controlled their migrations between seashore and interior, when to fish for salmon and sturgeon, when and where to find moose or beaver or birds' eggs. It dictated whether crops of berries and herbs would be heavy or light, and how difficult it would be to preserve them.

Finally, if necessity is the mother of invention, then weather mothered the snowshoe, toboggan, canoe, kayak, and umiak. So effective were these aboriginal solutions to weather's challenges, we still use most of them. Indeed, the canoe has become *the* Canadian icon.

Working the Land

Farmers are impacted by weather all year. Come spring, they have to get on the land as early as possible to plow, fertilize, and plant, each at the proper time. Should the ground be late thawing, or should there be heavy rains, they must wait. Sometimes they get on the land early, only to lose out later when cold and wet weather undoes their work.

Crop losses are a constant worry. A sudden downpour or hailstorm can flatten standing grain and hay, halting growth and making mechanical harvest almost impossible. Prolonged dry spells can stunt crops. One recent summer the soybean crop in our area was so short that, though the pods were full and plentiful, the machine's cutter rake could not be set low enough. It left the lower third of the crop in the field for the crows to harvest.

The ultimate cat-and-mouse weather game is haying. No wonder many farmers have switched to damp sileage harvested at the peak of nutrition and stored in concrete silos or under plastic instead of in barns. This method has the bonus of avoiding spontaneous combustion. Many a barn has gone up in flames as the temperature of fermenting hay ignited the rafters weeks after storage.

While dry weather is good for making hay, it's no good for growing it. Hay—traditionally timothy, clover, and alfalfa in our region—is a cool weather crop that does best during the showery weather of May and June. Of the early and late crops, spring hay is lusher and higher in protein. Prolonged dry weather lowers the yield of both.

Last year my dairy farmer neighbours, hoping for rain, let the July hay stand and stand. The standing hay, besides being short for the time of season, had a peaked look. When the rain finally came, the fields flushed green again. A few days later they were ready to cut.

For those who work the land, the return of rain after weeks of dry weather is a kind of reprieve, a sign that the universe is not out of kilter after all. It brings that peculiar satisfaction of going indoors after everything outdoors has been put in order, a sense of completion, of working with nature, not against her.

Today's bigger dairy operations have switched to crops more amenable to big machines working big fields, such as cattle corn and soybeans. Even so, weather still determines whether they will have a bountiful harvest with cheap winter fodder, or a meagre harvest with big bills for imported Western feed.

Farm Erosion

Agribusiness has a weather-related problem that old-time eastern dairy farmers seldom encountered—erosion. Hay ensures year-round sod cover, but monocultures of corn and soybean dispense with sod. Bare ground is worrisome for at least two reasons. The freeze-thaw cycle of our seesaw weather can kill fall-sown cover crops like winter rye. It can even damage deep-rooted perennial crops like alfalfa. Erosion is especially worrisome after fall plowing, or after harvesting wide-sown crops like cattle corn, potatoes, or soybeans. Even when protective stubble is left, bare soil is unavoidable, leaving the land vulnerable to sheet and gully erosion by wind or water.

Researchers in northern New Brunswick's potato belt found soil losses averaging forty-five tonnes per hectare. It's calculated that a loss of even a few millimetres of topsoil on twenty hectares can amount to over forty truckloads of topsoil a year.

Gardening

Gardeners, whether hobby or commercial types, whether they raise vegetables, flowers, herbs, or all three, must tiptoe around the weather. One year the crop is superb, the next it's poor. Too much rain after sowing, and the seeds may rot. Too little rain, and the smaller seeds like carrot and celery may not germinate without watering. Sudden rain after a fortnight of August drought, and the tomatoes may split and develop blossom end rot. A cool wet summer, and the cukes and corn may fail to ripen.

We learn to take what comes. "You win some and lose some," certainly applies to gardening.

Weather also dictates which crops will grow where, and how well. A novice can grow cool-weather stuff like spinach, onions, peas, and cabbage without much trouble. The same goes for cool-weather flowers like irises, peonies, and buttercups. Likewise perennial kitchen herbs like basil, thyme, chervil, mint, and parsley thrive with little care.

The trouble begins when we aspire to hollyhocks and delicate tea roses, to eggplant and watermelons, or even to tomatoes, peppers, and squash. Hardier new varieties have improved the odds of success, but some are still a gamble in the northeast. One needs all the tricks of the trade.

"Risk of frost in low-lying areas tonight...." Nothing panics the gardener more than those words. Rush to bring in the potted stuff! Close the cold frame windows! As the sun goes down, grab old sheets and drape them over the cukes and tomato plants!

"Mind the full moon in June," oldtime gardeners would say. For they knew that a cloudless night can mean frost, even after a warm day, especially inland, away from the moderating effect of the sea. But a full moon brings no frost when it's cloudy, for clouds hold in heat like a blanket. It's just that on clear nights when most frosts occur, a full moon is highly visible.

Orchardists and berry growers too fear late spring frosts. Fruit tree blossoms are delicate things. Even if the flowers escape, a wet, cool May and June discourages bees and other pollinators, resulting in scanty pollination. And early fall frosts can blight ripening fruit, including blueberries.

The *Canadian Global Almanac* in its weather section lists the 1-in-10 odds for a late or early frost in each province before a certain date. Charlottetown and Halifax can expect the last frost on May 27 and 28, respectively; for Corner Brook and four places in New Brunswick, it's June 10; for St. John's, June 24; and for central Labrador, June 27.

One would expect the date for the first fall frost to run roughly parallel. This does hold true for inland and/or northern sites like Fredericton (August 8, September 11, September 13); but not for places near the ocean. This is because the sea is at its warmest in the fall, so we get readings like September 19 for St. John's, September 30 for Halifax—and a cozy October 6 for Charlottetown, situated well away from the frigid Labrador Current on shallow Northumberland Strait, which is billed as the "warmest water north of the Carolinas."

If frost is the northeastern gardener's Enemy Number One, excess rain is surely Number Two. And the closer one lives to saltwater, the wetter things are likely to be. One day's downpour can wash out small seeds and uncover larger ones like peas and corn. Two weeks of wet, cool weather will rot them in the ground.

I've set potato seeds in balmy April sunshine, then waited until late May to see them sprout, the weather was so cool and wet. After a few years, you realize there's little to gain by pushing the season. Some years you might be lucky; most years the stuff you plant in mid-May will outgrow your April plantings.

Genuine cold-weather veggies like spinach, onions, and peas may prosper, but the rest will take their own sweet time. The same holds for most flowers. They dance to weather's tune.

It used to be that cold and wet were our main obstacles. In recent years heat and drought have been added to the gardener's headaches. While hot weather is essential for ripening corn, squash and tomatoes, too much of it can wither unwatered crops, stunt the growth of orchard fruits and berries, and encourage insect pests to breed in greater numbers.

Business

Store owners all agree that seasonal items need the stimulus of suitable weather to lure customers. This applies as much to tools and implements as it does to winter clothing. The winter of 2001 dumped more snow on the Maritimes than anyone had seen for a long while. Sales of snow shovels, roof rakes, and snowblowers blossomed. Many stores ran out, and even their suppliers were hard pressed to maintain stocks. A few weeks later, when rain and mild temperatures slicked roads and driveways with ice, they had a run on bagged sand and salt. Car washes also did a brisk business. As one operator put it, "When it's snowing people don't really care."

Many companies depend on specific weather conditions. A ski hill operator who fails to get heavy snow in December and January may go out of business; artificial snow just doesn't draw a crowd. Investors in breweries, ice cream makers, swimming pools, and dehumidifiers clearly have a major interest in hot weather. Conversely, those who hold stocks in natural gas and other fuels make their money in cold weather. So do those with holdings in snowmobile and snowblower companies, not to mention makers of cold remedies and facial tissues.

Here's a stock market tip! Buy when it's raining or overcast and prices are depressed. Sell in the sunshine when traders and investors feel better, buy more and doing so bid up the price.

—David Phillips, *Blame It on the Weather*

Natural disasters in one part of the world create a market for lumber and other building materials in another. In the fall of 2000, after widespread flooding and mudslides in Central America, the demand for Maritime lumber skyrocketed. Investors in agribusiness watch the weather constantly, even hiring forecasters to calculate the probabilities of crop success and failure.

A somewhat less obvious example is the insurance company. Weather-related damage can cost them a bundle unless they can prove "an act of God."

Transport Workers
Pilots

Of all those who work in moving people and goods from place to place—bus and taxi drivers, truckers, and train engineers—no professional heeds every hiccup of the elements as keenly as the aircraft pilot.

Pilots play weather roulette every working day. And none is more wary of the sky's changing moods than the bush pilot. Canadians pioneered this profession in the 1930s, carrying mail, supplies, and medical personnel to regions cut off from the outside world. They still do it.

Handling such small craft as the *Norseman* and the *Beaver*, they simply cannot ignore the weather. High winds, crosswinds, heavy snow, freezing rain, dense fog, choppy lake water, rotten ice—these are hazards which bush pilots face every day. Luckily, between the extremes of perfect flying weather and being grounded, there are infinite variations.

One variable which most of us never think twice about is atmospheric pressure. Skilled bush pilots take it very seriously. Aircraft respond to the controls better in dense, high-pressure air, for the same reason the Dead Sea's dense salt-rich water buoys a swimmer. Furthermore, one can see farther and sharper in cool dry air than in summer haze.

Migrating birds generally wait for a high-pressure system before setting out, often flying at night to enhance that advantage. During deep lows, even strong fliers like seagulls often rest on the ground.

Light, warm air, on the other hand, makes a plane work harder. It takes more fuel just to stay "afloat." Take-offs are longer and landings trickier, especially in gusty weather. For these reasons, veteran bush pilots like to take off early or late in the day, when the air is cooler and denser. Flying instructors also prefer those times; it boosts the margin of safety.

Bush pilots have another reason to mistrust warm air. It is the playground of updrafts and downdrafts. Those pretty marshmallow clouds we admire on sunny days are warm air cells. Each column of rising air displaces denser cool air, which plummets in a ring around it. This is harmless enough in normal weather. But where temperature gradients become violent, winds can develop. These vertical winds can fling a small plane hundreds of metres up or down. Often a pilot gets no warning except a spinning altimeter and perhaps a sensation of popping eardrums and of a brick in the stomach.

Commercial airline pilots in their big planes enjoy a greater margin of safety. Even so, strong headwinds are normally avoided because they increase fuel consumption. For the same reason, tailwinds are welcome. The most dangerous times are during take-off and landing, when strong turbulence can be fatal. The most dangerous condition is clear-air turbulence, so called because it displays no telltale warning clouds. The air behaves like standing waves in a whitewater river and it has dropped planes so suddenly that even belted passengers have been killed in mid-air. A low-flying aircraft may crash.

Smart pilots take no chances, especially in regions or seasons known for tropical storms or tornadoes, which are giant whirlpools of warm, moist air. Thunderheads are usually given a wide berth as well. Microbursts, a violent species of downdraft that can suddenly drop a plane 300m/1,000ft, lurk around them. Many airlines have installed costly laser detection equipment that gives the pilots enough warning to take evasive action. As a general safety precaution, most longer commercial airline routes try to fly above such disturbances, at 9,100m/30,000ft or higher.

In 1854 the U.S. clipper *Flying Cloud* sailed from New York to Boston via Cape Horn in 89 days, a record not broken until 1989 by a state-of-the-art yacht.

Seafarers

Next to pilots, few are so attuned to weather as those who, in the words of the Anglican *Book of Common Prayer*, "go down to the sea in ships, and occupy their business in great waters." Only 150 years ago, mariners were totally ruled by weather. Without a breeze, a sailing ship couldn't even leave port. Too much wind, and it risked foundering in heavy seas. In *Two Years*

Before the Mast Richard Dana Jr. describes how their square-rigged ship battled headwinds, blizzards, and mountainous black seas at the southern tip of South America for weeks, trying to round Cape Horn on their way to California.

Wind direction was also crucial to oldtime ships. Even today, no sailing vessel can head directly into the wind. The square-rigged ships of yesteryear sailed best "wing and wing," with the wind almost directly astern. This was because the sails were rigged crossways to the hull, which meant the yards from which the sails hung could be swivelled only so far to port or starboard before they brought up against the ropes that braced the masts and worked the canvas. However, with a brisk following wind they could foam along at a good clip. The so-called clipper ships (Scandinavian: *klippa* = cut), first built in the U.S. in 1845 and later in Britain, had a narrow hull that sacrificed cargo space for speed in order to compete in the American-Anglo trade with the Orient.

Their power came from a vast spread of canvas. With two or three dozen immense sails hung below horizontal wooden yards that could swivel halfway around masts as tall as church steeples, a full-rigged ship captured all the wind there was.

The idea behind the East Coast schooner was greater maneuverablity. Its fore-and-aft (front-and-back) rig deployed great bel-lying trapezoids of canvas which could be set slantwise to the breeze (close-hauled, sailors call it) so the vessel could sail on a zephyr and head almost directly into it. Sailing on the quarter, at around 45 degrees to the wind, they skimmed along like an iceboat. Unlike the iceboat, which because of the vector effect can actually sail faster than the wind that drives it, the schooner is slowed by friction. But it is still speedy. Nova Scotians still brag about their beloved *Bluenose*, which outsailed the New Englanders year after year in the schooner races of the 1930s. No wonder rum-runners loved the fore-and-aft rig.

In the glory days of commercial sailing and Banker voyages, many permutations of the square-rigged and fore-and-aft rigs graced the seas. They were the most beautiful wind machines the world has ever known—windmills included.

For all the changes in sailing craft, today's sailor craves the same things the old seadogs craved, namely clear sailing in the right direction with good visibility.

Commercial fishers are no different, except that they rely on powerful motors and can usually escape a building storm. To find out what's really in store, these people not only listen to up-to-date marine forecasts, they learn to read the early morning

She was a tiny full-rigged ship—perfect in every detail—a graceful and dainty little thing, as she picked her way out of the harbour with that clean wake which tells of perfect underwater lines.

—Alan Villiers, *The Set of the Sails*

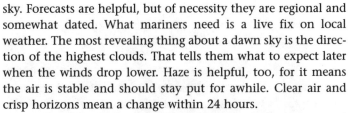

Sailors approaching land take note of the wind direction offshore and mentally adjust its direction to 90° off the coastline; they know it will shift to that direction as they go in.

sky. Forecasts are helpful, but of necessity they are regional and somewhat dated. What mariners need is a live fix on local weather. The most revealing thing about a dawn sky is the direction of the highest clouds. That tells them what to expect later when the winds drop lower. Haze is helpful, too, for it means the air is stable and should stay put for awhile. Clear air and crisp horizons mean a change within 24 hours.

Mares' tails and mackerel sky don't mean much, except good sailing for a day or two. The clouds which mariners respect most are those sullen, long, low, grey clouds that sneak up across a drab sky. They mean wind and rain. Another warning sign is sharp horns on a waxing or waning moon—that means new air coming in. Sharply twinkling stars mean the same thing, and if the stars hide the rain is near.

Veteran sailors also learn when to expect the wind to shift, and to what quarter. As old Westy Lyons of New England told weatherman Eric Sloane, "Any oldtimer knows that the wind blows like a pinwheel. If you want to locate any storm, just face the wind and point out with your right hand. You'll be pointing into the center of the pinwheel; at the nearest storm. You can follow the eye of a hurricane that way too." Sloane likely said, "Absolutely right, Westy—in the northern hemisphere, and exactly opposite in the southern."

Forest Workers

One day I was sawing down a large swamp maple for firewood. The air was calm. I expected no problem felling the tree where it would be easy to reach with the tractor. But just as I began the back cut, a breeze came up, causing the tree to lean in the opposite direction, pinching the bar of my power saw.

When this happens with a smaller tree one can often rig a push-pole or "samson" and overcome the wind's drag. But with a big tree in a stiff breeze, there was nothing I could do but go for help. This was urgent, for the tree had suddenly become a menace to any person or animal that came near. With a friend's tractor we were soon able to reverse the tree's lean and bring it down properly.

Likewise, if a gust of wind arrives just as a tree begins to fall, the tree may swivel and fall in another direction, perhaps injuring or killing someone.

Forest workers deal with such weather hazards almost daily. Other weather-related activities include tower lookout duty, building logging roads that will handle water runoff safely, and fighting forest fires (which increase with hot, dry weather).

Like farmers, enlightened forest managers also fear erosion, and take steps to prevent it. These steps include not logging on steep slopes, putting in runoff ditches or deflector logs to channel excess water harmlessly across roads, and installing culverts and bridges in ways that minimize siltation of streams.

Silviculture operations such as thinning and planting must also take account of wind and weather. When crowded stands are opened too fast on shallow or wet soils they often blow over, especially on south-facing slopes where high winds hit hardest. Old trees are especially vulnerable, whether thinned or not.

On a smaller scale, tree planters must watch out for frost-action. If a seedling is planted in raw mineral soil, frost action is almost sure to heave it out before next spring. Planters are trained to set their seedlings in organic matter where possible. This is especially important with baby trees grown in greenhouses in smooth-sided containers.

Trappers

In a sense the trapper, whatever we post-moderns may think of his seemingly cruel trade, is the best attuned to the niceties of weather, for he lives out in it a good part of the year. My father trapped to make ends meet during the Great Depression, so I know. Thousands of people still trap in eastern Canada. It is not a sport, but a strictly regulated profession with a code of ethics, licences, seasons, and bag limits.

Early snow, if not so deep as to hamper movement, makes trapping so much easier. Successful trappers don't just go out and set traps. Several weeks before the season, they scout out a possible territory. They do this by reading animal signs in the snow. Each furbearer leaves a distinctive signature and follows a different routine. The trapper checks for browsed twigs, recent droppings (scat), and recent kills. Knowing what the quarry eats, when and where it travels and sleeps, the woodsman locates a trapline.

Usually the ground is bare when he does this, but his task is made much simpler when snow comes early. Then his trained eye can decipher from the snow-covered ground what he needs to know.

Indoor Workers

So much for outdoor workers; what about those who work indoors? The late American weathercaster and writer-artist Eric Sloane maintained that most great works of Western writing and painting have been done in February. Well, it makes sense. Most creative work is done indoors, and most of us are indoors then. But there's more to it. Many of us feel a hibernation urge as winter closes in, a desire to hole up in a snug place with a good fire and plenty of food and fuel, the stormier the better. I call it the Groundhog Syndrome.

Certainly long winter nights open the possibility of doing things we were too busy to tackle in the long days of summer. People who hardly stop for meals in summer suddenly find the time to scrape and varnish an old table, carve a duck, write a long letter or a short novel, paint a picture, take up the sax. Hobby store owners do most of their business from November to March.

Office bosses would argue that the negative aspects of winter weather far outweigh the positive. Absenteeism due to sickness rises sharply. There are delays due to storms and accidents, flights are cancelled, and deliveries are late. Tempers flare because people are under greater stress, both at work and at home. School boards and principals have to contend with up to a week of lost time due to closures—and then make it up in June.

Plasterers and carpenters have different weather problems. A perfect wallpapering job can come unstuck after weeks of wet weather. Doors hung in dry winter air may bind when damp summer air returns unless allowances are made. Drawers that slid perfectly in sunshine may stick in rain. Perfectly mitred corners can widen as temperature and humidity change. Knowing how to foil the weather is part of the skilled worker's craft.

<aside>
Recent *Maclean's* magazine polls consistently rank Newfoundlanders highest in the category of Active Sex Life, which Newfoundlanders credit to chronic unemployment and long winters.
</aside>

Weather

& Health

In January of 1999 Michel Trudeau, youngest son of Canada's former prime minister Pierre Elliott Trudeau and Margaret Trudeau, was hurtling on skis down a snowy mountainside in the Canadian Rockies. Above him soared a pristine wall of white, below him lay a jade-green lake. As he swooped down the slope in a cloud of snow, the wall of white shuddered and began to move. By the time he heard the rumble it was too late. The avalanche engulfed him and bore him into the lake. Despite a thorough search that summer, his body was not found.

Snow in all its forms can be lovely, but it can also be deadly. In Canada, snow and cold have traditionally caused more deaths than any other weather phenomenon. Most of these deaths aren't spectacular, but caused by hypothermia. The body becomes too cold to function. Every winter in our cities, street people die this slow death. In February of 2001, with temperatures near -20°C/-4°F, a teenager who was camping in a forest cabin with some friends left to walk at night to a nearby car, lost his way, and perished in deep snow.

Those climate victims weren't trying to prove anything—unlike Leonidas Hubbard, the bored American socialite who blithely tackled the Labrador wilderness in 1903. Hubbard went up the wrong river, ran out of food and spent his last days shivering in a tent in the snow, eating his moccasins and scribbling last thoughts in a journal. There's Sir John Franklin, who disappeared in 1845 trying to discover a northwest sea passage across northern Canada. Arctic and Antarctic explorers—Scott, Rasmussen, Peary, and others—have been driven almost to despair by snowstorms; and mountain climbers, holed up in a miserable pup tent within sight of their goal, have cursed the fresh snow that hid the crevasses.

Snow really hampers movement once it gets deeper than half a metre. The winter of 2000-2001 dumped so much of it that at times I had to snowshoe to the mailbox and back. It became necessary to clear the front steps, garage door, and patio almost every week. It didn't bother me, but I thought of all the out-of-shape men my age and younger who have suffered heart attacks while shovelling snow.

Canada, like Siberia, is a cold country. While we don't get a lot of droughts and floods—at least we didn't used to—we do get Arctic temperatures. Usually it's our neighbours to the south

Whoever wishes to pursue properly the science of medicine, must proceed thus. First he ought to consider what effects each season of the year can produce...The next point is the hot winds and the cold...but also those which are peculiar to each particular region.

—Hippocrates, *Airs, Waters, Places* c. 400 B.C.

who swelter in heat waves that kill the elderly and the frail with dehydration and sunstroke. Not only do we die of hypothermia, we commonly freeze ours ears, nose, fingers and toes, and sometimes lose them.

Wind Chill

It's not just low temperature that maims and kills, but the wind that often goes with it. Wind penetrates our clothing and removes warm air faster than it can be replaced. Physicians now recognize that our bare skin is normally encased in an insulating and moisture-conserving film of air 4-8mm (1/6-1/3in) thick. We carry our weather around with us. Cold wind strips this film away, causing us to shiver if we lose more than we can replenish. To describe the combined effects of low temperature and wind, the term "wind chill" was invented. Wind chill is an index of heat loss and cold injury that combines temperature and wind velocity, neither of which alone gives a realistic picture. To calculate it, estimate wind speed in kilometres per hour, divide by two, put a minus symbol in front, and add the result to the Celsius temperature. Thus, if wind speed is 20 kmh and temperature is -15°C, the wind chill temperature is -10 plus -15, that is -25°C. Thus the victim would feel almost twice as cold as a thermometer would indicate.

Paul Siple, an American geographer and polar explorer, coined the phrase *wind chill* in 1939.

Weather has other effects. Cold, damp weather inflames the joints and sinuses of arthritis and asthma sufferers. Those who can afford it escape to hotter, drier climes for a season or for good. A lot more Canadians would stay away if the prospect of losing their Medicare benefits didn't draw them back.

No, Canadians seldom forget that the North Pole is just over the hill. If we get caught out in a blizzard in a stuck vehicle, or if we've been drinking and rowdy in a northern bar at Christmas and the police drop us outside town to cool off, we may never make it home. Or if, like the jilted farmer in Sinclair Ross's short story "The Painted Door," the victim goes out in the cold not caring any more, he may die. They say hypothermia is not a bad way to go. Once the shivering has passed, and the pain in hands and feet and face subsides, the victim is said to bask in a feeling of great warmth and comfort. I've never gone so far, but I've felt the onset.

It happened while I was attending Mount Allison University in Sackville, NB in the mid-1960s. One bitter winter night I set out on my motorcycle across Sackville's Tantramar Marsh to see a new art exhibit in Amherst in which a painting of mine was on display.

Since it was only an 18km/11mi trip, I took no special pains to dress for it. Besides, the journey there, with sunset warmth still lingering in the blue air, wasn't all that cold. Two hours later the mercury had dropped dramatically. A gentle breeze had sprung up, not very cold, but a motorcycle doing even 50kmh/30mph generates a wind chill that can penetrate the warmest clothing.

When I came out of that warm gallery into the blue-black night glittering with stars, I shivered involuntarily. My two glasses of wine didn't help either. Kick-starting the motorcycle, I wheeled onto the street and set out. In the lee of the buildings it was okay, but out on the marsh the wind cut like a razor. A pang of fear came over me. Should I go on? Would I even make it? I decided to risk it.

The trip took me over an hour. Speeding up would have frozen my face and gloved hands; slowing down prolonged the cold unbearably. My body felt encased in ice. There was no relief. As my hands grew too numb to feel the handlebars, it seemed likely that I could fall off the bike. I clung to it as a drowning person clings to a paddle. After a while, the machine seemed to be thinking for me, taking me home.

At last it climbed the long hill to our apartment outside town. Long after midnight, I pried my body off the seat, unbent my legs, forced my feet down to the squeaking snow and clumped up the three wooden steps into our kitchen. At first the warmth hurt. Then it became a hunger. When I crawled into bed, my wife shuddered at the cold of my skin. It took me two hours to stop shivering and fall asleep. Next day my hands and face had a sunburned feel. I had narrowly escaped frostbite or worse. After that, I always dressed appropriately.

Once one freezes a body part, it's more likely to freeze again. For instance, as a young boy my eartips froze. Now they freeze at slightly higher temperatures. They warn me by tingling. A mild frostbite like that is called frostnip. Frostnip can be treated

It is said that the Inuit deliberately exposed young children to frostnip to sensitize them against more serious injury.

Note: Never rub snow or ice on frozen flesh; this only prolongs the cold and can abrade the skin.

on the spot by blowing one's breath on the affected part, placing nipped fingers in one's armpit, or holding a bare hand over the affected part.

If deep frostbite is suspected, get the victim to a warm (not hot) place, and summon medical help. Third or fourth degree frostbite is a medical emergency which can lead to gangrene and amputation.

Winter Driving Precautions

Let's assume it's a blizzard, and you're stranded on the highway, but you've managed to contact a towing service via cell phone or another driver. There's still no guarantee they'll reach you soon. Check whether the vehicle carries basic survival equipment. A winter road kit should contain:

- Working cell phone
- Sturdy shovel (small and flat enough to reach under a car; a metal garden spade is excellent)
- Bear-claws or a bag of hen grit (for getting unstuck—far better on ice than salt or sand)
- Sleeping bag(s) or blanket(s)
- Wind suit, mitts, and reflective vest(s) in case you need to change a tire or walk for help at night.
- Sealed, freezable emergency rations (hard candy, chocolate, raisins)
- Wooden strike-anywhere matches in an airtight container (kept dry with silica gel or salt)
- Candle(s)
- Flares or reflector
- First aid kit

Assuming your vehicle has most or all of these (a big assumption), the battle is half won. While awaiting rescue, here are some things to do and not to do:

- Stay with the car; it's your best shelter.
- Ensure exhaust pipe is free of snow so deadly carbon monoxide won't seep in.
- Run engine for ten minutes every hour. This should prevent hypothermia and charge the battery, while saving gas.
- When running car, open one window a bit for fresh air.
- Place a warning light or flare to make sure vehicle can be

seen. Snowplows have collided with unmarked stranded cars. Even a scarf tied to the aerial will help.

- Don't drink alcohol to keep warm; it will chill you.
- Bundle up; cover your head especially—up to 70% of heat loss occurs through the scalp.

In the days before sunglasses became fashionable, snowblindness was another winter hazard. The highest incidence was among sealers. The glare from horizon-to-horizon ice on a sunny March day is punishing. Every year several men on each ship would lose their sight for a few days the first week. Folk remedies like poultices of raw potato or tea leaves seemed to help.

Eons ago, the Inuit solved this problem by wearing slotted wooden eye shades. Smart sealers wore blue-tinted goggles. My father wore a pair on his winter trapping journeys for years. Sunglasses have appealed to me ever since.

They say the first humans lived in the warm womb of subtropical Africa, a Garden of Eden if ever there was one. But it wasn't in Africa that we learned to cope with snow and cold. We learned that in post-glacial Europe and North America. The ice ages pruned the weak, honed our hunting skills and shaped our view of nature. Today those skills and attitudes have been subsumed into modern life.

Odours

One of the distressing effects of bad weather is unpleasant smells. Our noses get unhappy, and so do we. The human sense which has suffered most from civilization is our sense of smell. Perhaps it was never that good, certainly not compared to that

of any wild mammal. A pet dog's sense of smell is about four hundred times better than yours or mine. (That's why our dogs get so much more joy from rural walks than we do.) Our noses are too short and our brains devote too little circuitry to smell. Even so, we notice more odours during rainy or snowy weather. This is because of lower atmospheric pressure which allows bad smells to escape the cat's litter box, the refrigerator, and such.

Then there's "muggy weather," when a southerly wind settles in for days on end, bringing that hot, oppressive kind of weather that makes a single bed sheet feel as heavy as a quilt, when clothing sticks to your back, hair is lank, and nothing will dry. In such weather tempers flare, children get irritable, sneakers smell bad, and nothing seems to go right. "Gulf of Mexico air" I call it.

I used to think it was all in my head. Then I discovered I wasn't alone. Sultry weather really does get under people's skins. It has been noticed all over the world.

In Europe the south wind has been blamed for headaches, sluggishness, and dullness of hearing and vision; and the north wind, called *Boreas*, has been blamed for coughs, throat infections, and even constipation.

Vincent Van Gogh, painting non-stop outdoors in the south of France, came unhinged partly from the relentless buffeting of the cold dry *mistral* ("masterly") from the north and by the hot damp *sirocco* from Africa. He had to weight his easel down with stones and clamp his hat down tight. But I hadn't realized until reading Jan DeBlieu's *Winds* that there were so many other ill winds. She lists the *bora* ("northerly") of the Dalmatian coast; the horrible *khamsin* ("fifty-day") of Egypt; Turkey's *samiel* and North Africa's *simoom* (both meaning "poison"); the *melteme* ("bad-tempered"), a northeaster of the Aegean; the tornadic *yamo* ("in-the-body") of Uganda; the violent *messar-ifoullousen* ("fowl-plucking") southeaster of Morocco. Our word "hurricane" is from the Carib-Indian word *huracan* meaning "evil wind."

But there are some good winds: the *trauben-kocher* ("grape-ripener") of Switzerland, the dusty but welcome Saharan northeaster *harmattan* ("doctor") of Africa's humid Guinea coast, and the mild south wind *shawondasee* ("lazy wind") of the eastern Algonquins.

Goosebumps happen when a chill or a fright causes the fine hair on our bodies to rise in unison to preserve the skin's film of insulating air.

Then into the night came the sound of the wind, beginning like a gasp released out of a pin in the ground, from under a wave or through a hole in the ink-black sky.

—Susan Kerslake, *Middlewatch*, 1976

Cold and dry winds bother us more than cool and humid winds; hot and humid winds are the worst of all. As Hippocrates noted, people suffer in various ways. Some feel "woolly-headed," unable to think. Some don't want to touch anyone or be touched. Small problems loom large, things look hopeless, there's a "what's-the-use" feeling, people feel lazy. In short, hot southerly winds bring on symptoms of depression. Police say the crime rate rises too.

On the other hand, we all know the lift that comes with an influx of cool, crisp westerly wind that shoves the stale locker-room stuff out to sea amid protestations of lightning and thunder. When that happens we don't need a barometer to tell us the pressure has changed. If you check, though, you'll see it climb as steadily as your spirits.

Animals feel the lift too. Kittens climb drapes and cats chase their tails. Horses gallop and kick up their heels. Cows that had lolled around the pasture for days suddenly gambol like spring lambs. Seagulls that sat about while the pressure was low now unlimber their wings and climb into the air again. In this they resemble human swimmers who prefer salt water to fresh because it's more buoyant. Suddenly we too like playing outdoors, doing things, and going places. All because of a change in wind direction.

Wrinkles

Just as the elements soften and abrade exposed wood and brick and stone, they also age our skin. Anyone past the age of forty knows something of this. Women in particular feel the stigma. It wouldn't be so bad if we lived in the Orient, where a crinkled face is (or was) honoured as a sign of maturity and wisdom. But the West prizes youthfulness over age, and furrowed skin can be distressing. The only real facial weathering we honour is the bronzed, leathery look that hardened gardeners, surfers, rock climbers, and northwoods guides get for free.

Wrinkles, unlike hidden aches and pains, are undeniable outward signs of aging, and weather is the prime cause. Moreover, ultraviolet radiation broils our skin unless we grease it before venturing into the hot sun. We powder or paint the wrinkles and crow's feet over, mud-pack them into submission. A thousand products and services promise to conquer them, to prolong that smooth-skinned look for men and women both. And if all else fails, there is always plastic surgery.

A Thirst for Light

Psychiatrists have long known that weather influences our feelings. Low pressure can actually trigger mild depression. Not clinical depression—which is rooted in genes and body chemistry and life situations—but "Ah-what's-the-use?" depression, the kind that hobbles everyone from time to time.

Who hasn't experienced the midwinter blahs? It hits people who have no history of real depression. While a few people get depressed from too much sunshine, most crave it, especially in winter. Winter creates a hemmed-in feeling, a sense of imprisonment. The aged and infirm feel it most. Afraid to venture onto the icy sidewalks, perhaps unable to drive a car, they stay indoors. Those who can afford it head south to Bermuda or Cuba or Florida for a short or long holiday in the sun. Those who can't afford it stay and tough it out.

The symptoms of this midwinter malaise are so common and so predictable that psychologists have a label for it: Seasonal Affective Disorder, or Seasonal Adjustment Disorder, SAD for short. SAD hits hardest in December, when the nights are longest and the cold weather begins to force us indoors. But it's not so much low temperatures as lack of sunlight that drags us down. Humans, like begonias and chickens, need a minimum of light each day in order to thrive.

Most people spend more than 23 hours of their 24-hour day indoors, where illumination rarely exceeds twilight levels.

Farmers and horticulturalists have known this for a long time. When I kept hens, it was necessary to have a light in their house from November through April, or they would stop laying no matter how much protein they were fed. But when I gave them 16 hours of light a day, they gave us eggs all winter. They didn't live as long, but they produced. Likewise, growers of hot-house cukes, tomatoes and begonias control the blooming times by manipulating the ratio of light and darkness.

What about humans? If anything, humans get far too much light—and therefore too little sleep. Ever since the Nova Scotian Abraham Gesner invented and popularized coal oil (kerosene) in the mid-nineteenth century, we haven't had to go to bed so early. Thomas Edison patented the first successful incandescent light bulb in 1879, and by 1882 he had 85 customers paying for electric lighting in New York.

But electric lights are no substitute for the sunshine. NASA and others involved in prolonged space flight have naturally studied the subject in depth, since light triggers vital body rhythms, and astronauts lose track of day and night. They find that when light hits our retinas it sends messages to the "body clock" located in the hypothalamic centre deep in our midbrains. When this clock is properly set, our day/night activity pattern matches our work schedule and lifestyle and we "feel better."

Scientists even claim that daylight acts as an essential body "food" like air, food, and water, and that most of us are daylight-deprived. As one might expect, the most malnourished are: (a) northerners (whose daylight hours are seasonally shorter); (b) those who live in cloudy or overcast areas where "dark" days are common (e.g., St. John's); (c) homebound persons who cannot easily get outside; and (d) those who work irregular hours or night shifts. Weather is implicated in all but the fourth. The second category pretty well describes our region in winter.

The foremost symptom of daylight deprivation, as noted above, is mild depression, which is marked by low energy, moodiness, social withdrawal, and lack of interest in sex. Daylight-starved people also have trouble falling asleep or waking at desired times. They often feel drowsy or tired during the day, with a loss of alertness and performance. They have trouble adjusting to irregular travel or work schedules. They often crave carbohydrates and are apt to gain weight.

How much daylight do we need to set our clocks right? Judging by the exposure our outdoorsy ancestors got, a great deal. Studies show that most of our activities take place in light levels dimmer than sunrise (i.e., half the sun showing) on a clear spring day. Even a well-lit drafting room is only marginally brighter. By contrast, one hour after sunrise the amount of light falling on the retina has climbed ten-thousand-fold. By noon it is 81,000 lux, declining to 10,000 at 5:10 PM, 5,000 at 5:30 PM and 750 at sunset. If that is any guide, we are deprived indeed!

Another factor in SAD may be the decline in vitamin D as we cover up for winter (bare skin can synthesize vitamin D under the stimulus of sunlight).

All this raises the question of how the oldtime Inuit living above the Arctic Circle managed to survive six sunless months with whole families shut up for days in tiny igloos with nothing but soapstone lamps, starlight, and the aurora borealis to see by. It was a marvel of resourcefulness. They played a lot of games, ate a lot, slept a lot, talked and sang a lot, did soapstone carving and scrimshaw, made and mended clothes and tools, had a lot of sex, argued and fought.

No doubt the closeness of the Inuit family, steeped in the spirit world, also helped banish dark thoughts. Perhaps in such a light-deprived environment—almost impossible for we floodlit post-moderns to comprehend—the flickering yellow flame and dancing purple shadows of a seal oil lamp became as good as sunlight. And of course they could look forward to their six-month day.

"April is the cruellest month," wrote T.S. Eliot in "The Waste Land," meaning the season when winter is not quite over but our resources are running low. Certainly late winter was cruel for the pioneers. Even First Nations people had a hard time. Food ran low, travel was risky, and hunting and fishing were often impossible. Game was scarce, underfed, and wary, while sturgeon, gaspereau and shad had not yet returned to spawn. The natives' response was to sit it out, performing only the bare necessities, fasting if necessary. Many starved.

Early European settlers tried to lay in surplus food and fuel the fall before. They disparaged the aboriginals for improvidence and "laziness." The fact is, prolonged fasting is an effective birth control strategy, one which kept the aboriginal population in balance with its food supplies.

Despite their scorn for the natives' "improvidence," our forebears did their share of unintended fasting too. Visit any pioneer cemetery and observe the months when most deaths occurred, especially those of women, children, and the elderly. Often it was March and April. Unlike their native neighbours before contact, they were beset by Old World maladies such as consumption (tuberculosis), diphtheria, and whooping cough. Winter confinement in small, badly-ventilated rooms spread those diseases. A TB-ridden grandfather coughing and spitting by the stove—often the sole source of heat—and eating from the same dishes often infected the weaker family members.

Immune systems were overtaxed, chills brought on pneumonia, falls caused broken bones, axe and knife cuts led to blood poisoning. Half of all babies died in their first year. Pasteur and Lister had not yet demonstrated the existence of germs.

Without antiseptics to prevent infection, "childbed fever" was a death sentence for many women. Counting infant mortality, the average pioneer life span was well under fifty years.

In less than a century all that changed, at least in the "developed" world. Thanks to safer drinking water, better food storage, antibiotics, education, and improved medical care, parents don't need to have seven babies in order to raise three to adulthood.

Today, of course, we have smog, vehicle mishaps due to ice, and heart attacks from shovelling snow. Winter gets most of the blame now. Though smog can kill at any time of year, it's worst when smoggy cold air gets trapped under stable warm air—a "temperature inversion."

Another hazard is the well-insulated, well-sealed modern house. I have a neighbour whose home is so airtight that the fireplace won't draw until a window or door is opened. So little fresh air gets into such homes that microscopic dust accumulates, resulting in asthma attacks and a proliferation of invisible dust mites.

Got the Flu

While everyone gets more colds and pneumonia in the winter months, most people take it in stride. As Dr. John Olds of Twillingate, NF, told a cold-stuffed uncle of mine who was seeking relief, "Well, Don, if you take nothing, it'll last two weeks; and if you take something, it'll last a fortnight." It's not the colds we worry about; it's the flu.

A few years ago *The New Yorker* published a cartoon in which a small chicken sprints across the hen yard yelling, "The flu is coming! The flu is coming!" The joke is, it's no joke. Every fall and winter, worldwide, millions of people of all ages come down for a week or a month or more with fever, headaches, muscle pains and general malaise. Influenza, grippe, call it what you will—the flu has struck again. People call in sick. Hospitals and other public places are shut down until the emergency passes. Some people, especially the old and infirm, develop pneumonia and die if they haven't had a flu shot in time. But most get better after a few weeks and forget the episode until next year's "flu season" rolls around.

Is there really such a season? No one knows. Just because flu usually hits in the fall does not necessarily mean the season is the cause. It may have more to do with people's seasonal habits than anything else: back from worldwide vacations, back to work, back to school, congregating for sports, concerts, church, parent-teacher meetings, clubs, bowling. In short, back to lots of indoor mingling, with lots of chances to trade germs.

However, fall and winter weather tax our immune systems. Chills and wet, the stress of driving in snow and ice, the renewed worries over paying college tuition fees and big heating bills, the steady loss of daylight from late September through late December; all these things take their toll and leave us less able to fight off germs, especially a microbe as versatile and efficient as the flu virus.

There is another factor that has only recently come to light—the virus's connection with wildlife, specifically waterfowl. Most people think of ducks as lovable, comic figures. We may have to rethink that.

In the late 1990s, virus experts uncovered a link between waterfowl and human flu. They found that waterfowl are the "reservoir" for influenza. Ducks, it turns out, harbour most of the known subtypes of the disease—but without apparent ill effect to themselves. Ducks excrete the viruses in their feces, spreading them through land and water ecosystems and to certain mammals. Fecal sampling shows that in September up to thirty percent of all Canadian wild ducks migrating south have avian flu, and by then have infected most lakes with it. Down south they repeat the process. Here then is a connection with winter flu outbreaks in the ducks' northern and southern habitats.

Virologists believe humans likely got the first influenza virus from wild ducks via pigs.

However, not all mammals get the flu. Among those that do are horses, ferrets, seals and pigs. And humans, of course. Once we get it, we do a good job spreading it among ourselves by coughing and sneezing and jetting around the globe.

Although waterfowl provide a year-round reservoir, they don't infect us directly. Avian flu viruses need an intermediate host, an animal with the same influenza receptors humans have. The only known species with receptors for both avian and human flu is the pig. This adds new meaning to the old British saying: "Dogs look up to you, cats look down on you—but pigs is equal!"

Jeffery Taubenberger of the U.S. Armed Forces Institute of Pathology, an expert in recovering genetic information from preserved tissue samples, has advanced the theory that the world's periodic flu epidemics may come from mild viruses mutating into killers as they move from duck to pig to human. It is a plausible and frightening theory. They know, for example, that the Hong Kong flu of 1968-69, which killed thirty-three thousand people, consisted of seven genes from an everyday human virus and one from a duck. The eight genes had shuffled their RNA inside a pig to create a deadly hybrid. The Asian flu of 1957, which killed seventy thousand Americans, may have developed in the same way.

To counter such threats, each autumn researchers test people worldwide for new flu strains. Racing against time—it takes six months to develop a new vaccine—they painstakingly identify and isolate the new bugs, inject them into millions of hen eggs, harvest the resultant cultures and have them ready for the next flu season. The program saves thousands of people each year, notably seniors, children, and invalids. Not many younger people bother with the vaccination. So far they haven't needed to.

But that could change. A virologist's worst nightmare is the recurrence of a supervirus like the so-called Spanish flu of 1918. Starting between September and November, it killed twenty to forty million people that fall and winter. What made the pandemic so unusual was that it was better at killing the young than the old. No one knows why.

What is known is that vast troop movements in the closing months of World War One helped spread the disease. The flu actually started in America, not Spain. The first known case was diagnosed March 4, 1918 at Camp Funston in Kansas. By April it had spread to most U.S. cities, whence it was carried to Europe by the hundreds of thousands of American soldiers crossing the Atlantic to fight in the spring offensives. The spring flu was serious, but not catastrophic until August, when an extremely virulent form emerged simultaneously in America and Europe, spreading wherever Allied soldiers went, including Africa, Russia, India, and New Zealand. Cold wet weather plus wartime shortages and general exhaustion only made matters worse.

Malaria in the Maritimes?

Among the most worrisome climate developments of recent years is the northward spread of tropical diseases attributed to global warming. In August 2000 *Scientific American*, in a major article on the subject, cited increases in cholera, dengue fever, yellow fever, and malaria in regions outside their normal range. In the summer of 1999, for instance, New York suffered its first deaths from West Nile Virus, a disease spread by mosquitoes feeding on infected birds. Since millions of songbirds migrate thousands of kilometres twice a year to nest and breed, and since mosquitoes are everywhere, the potential for spread is great.

Warmer weather allows mosquitoes to hatch more broods per breeding season, which means higher numbers and lower predator success. Likewise, shorter winters increase the survival rate and allow them to thrive farther north and at higher elevations. This also applies to other insects that carry disease, among them

Northern winter temperatures usually kill overwintering mosquito pupae, larvae and adults, thereby confining the insect to more temperate latitudes.

In August 2001, a dead crow tested positive for the mosquito-borne West Nile Virus in Ontario, the first in Canada.

wood ticks that transmit Lyme disease. Killer bees that have been moving north from the southern states may now extend their range into Canada. But mosquitoes are the main concern, because they interact with birds and birds travel with the seasons.

Come to think of it, I've been noticing more mosquitoes in recent summers. Neighbours who live closer to the saltmarsh have always complained of them. But we've always counted ourselves lucky to have so few, even in peak season. Since 1998, however, they have burgeoned, especially when rain follows a spell of hot, dry weather.

On January 30, 2001, during a thaw after a spell of severely cold weather, a mosquito stung my wife on the finger while she sat watching TV. Naturally, she didn't expect that, as it had never happened in winter before. We keep a cool house, heating only those rooms we need, yet mosquitoes now show up in every month of the year, usually in the bathroom and shower room, and sometimes in the upstairs bedrooms. Something is going on.

Dry weather and falling reservoirs due to global warming help concentrate disease organisms into smaller volumes of water, thereby further raising the risk of infection or poisoning.

Conversely, widespread flooding contaminates drinking water by washing sewage and farm fertilizer into wells and ground water. When torrential rains from a warmer-than-usual Indian Ocean soaked the Horn of Africa in 1997 and 1998, they set off epidemics of cholera, malaria, and Rift Valley fever which killed thousands of people and cattle. The wholesale felling of forests, which retard runoff, worsens flooding. The outbreak of *E. coli* poisoning in Walkerton, Ontario in the summer of 2000, which killed seven people and sickened two thousand, was due to flooding and careless testing. St. John's had a similar scare in July, 2001.

There you have it—frozen extremities, hypothermia, weathered skin, depression, colds, heat stroke, headaches, heart attacks due to over-exertion, flu and other respiratory ills, contaminated drinking water, tropical diseases—these are just some of the effects of weather on body and mind.

Weather

& Clothing

One of the proudest moments of my childhood came when my parents bought me a pair of real sealskin mukluks. The boots consisted of uppers made from single wrap-around pieces of hide topped by leather drawstrings, and skin moccasins sewn to the shank with tucks at heel and toe to fit a child's foot.

At first these boots crackled when I walked in cold weather, but once greased they were silent. Every seam was tightly sewn with animal gut and waterproof. To stay waterproof the whole boot needed to be greased often with warm seal fat, which hardened on contact with the cold. The ripe smell made the dogs follow me about, but I didn't mind because my feet stayed dry. The thin soles made it painful to walk on anything but snow, and the boots wore quickly in contact with gravel. My father, who knew how to work leather and who mended all our footgear, repaired them once or twice, then gave up. The next winter my feet were too big to fit them anyway. But I kept them for a while because they made me feel native.

The young naturalist Charles Darwin was astounded during his 1831-36 world cruise voyage aboard H.M.S. *Beagle* to see nearly naked Tierra del Fuegans apparently warm while he and his shipmates shivered in greatcoats. When the crew finally did

The price of a suit of long underwear in 1910 was 96 cents. In 1993 it was $24.99.
—David Phillips, *The Day Niagara Falls Ran Dry!*

STRETCHED BEAVER PELT

manage to entice them near the campfire, the natives sweated profusely. In fact, fire so frightened the inhabitants of the "Land of Fire" that they preferred the cold.

This was unusual. Perhaps they had lost the use of fire. Certainly no northern aboriginal went naked in winter. Warm clothing was essential, and those who failed to provide it perished. The ancient Dorset and other east coast peoples wrapped themselves in the skins of animals. Caribou was preferred because the hairs are hollow, a natural insulator. Animal hides were worn fur-side out in summer and reversed in winter.

In winter the standard upper garment for males and females was a hooded parka. The lower garment for men was pants; women wore culottes. Both wore leather mittens and knee-length watertight boots made from the hides of seal, beluga, or caribou. Clothes were decorated with hides of contrasting colours signifying the wearer's age and gender. Undergarments were made from furs and soft materials such as eider-duck skins. The coastal Innu of Labrador and the Beothuk of Newfoundland also dressed mainly in caribou and seal skins. Ethnographer Ingeborg Marshall stated that to properly outfit one adult Beothuk for the long island winter took six caribou hides.

Farther south, where a variety of forest animals abounded, the Algonquian peoples dressed mainly in the skins of white-tailed deer (which in winter also have hollow hair), beaver, and moose. The Mi'kmaq of the Maritimes and northern Maine made a sturdy boot by skinning the heel joint of a moose's hind leg more or less in one piece, shortening and sewing the front to fit. Settlers adopted the idea, added laces in front, and called it a larrigan.

MAKING A LARRIGAN

Fashion vs Weather

During the last two world wars groups of East Coast women commonly gathered to knit finger mitts. When they had enough pairs, they sent them overseas to the men in the trenches. The finger mitt is a hybrid glove-mitten with a separate index finger on one hand or both. It allows the wearer to pull a trigger or bait a cod hook in cold weather without removing the glove. It's convenient and warmer than cut-off finger gloves because the index finger can be poked into the main part of the mitt. The idea is said to have originated in Newfoundland. A modern version is the flap glove with a velcro fastener to allow the wearer to cover or uncover the fingertips at will.

From prehistory until quite recently, most clothing was made at home, and weather and occupation dictated what you wore. Almost every household kept sheep, and nearly every kitchen had a spinning wheel. Today most clothing is factory-made and weather plays second fiddle to fashion and its lackey advertising.

FINGER MITT

Fashion was the least of the worries of Acadians and Britons newly arrived in North America. One of their first tasks was to make new clothes before the garments on their backs wore out. As a result, most of the new clothes were utilitarian. Utilitarian and drab. All the tried and true Old World dyes, reds and purples and blues, were far away across the heaving Atlantic and cost a king's ransom to import. Except for the odd treasured dress or bonnet kept in a trunk for weddings and holy days, the newcomers wore mostly greys and browns and black. Homemakers had to relearn the art of making colourfast dyes using New World plant and animal substances. In time, with patient experimentation and by copying the natives, they found substitutes.

That all changed with the Industrial Revolution. In Britain and Europe factories sprang up like mushrooms after rain. Soon the mechanical loom, knitting machine, and garment factory made it possible to churn out broadcloth and clothing for every need in every colour, putting traditional home-based spinners, weavers, and knitters out of work.

One of the most popular productions of the mechanical loom was long underwear. Buttoned up the front, with a trapdoor in

STANFIELD'S KNITTING MILL, TRURO, NS (c. 1890)

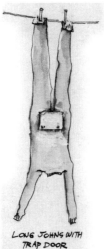

LONG JOHNS WITH
TRAP DOOR

the back, it became the standard for rural men and boys for several generations. The first were made of wool, which felt scratchy but absorbed moisture. Then cotton became more popular. One could buy a thin suit for summer and a fleece-lined suit for winter.

Jokes about long underwear were legion. Some oldtimer bachelors wore their one suit of Penman's or Stanfield's year-round. In summer they might roll up the sleeves and open the top two buttons for ventilation. Come fall they would button up again, covering the vee of sunburned skin at the neck. The incidence of skin cancer was low.

It didn't take long for First Nations people to adopt paleface attire, especially when trading a few beaver pelts could outfit one's whole family. Given the choice of killing a moose and skinning it, of tanning and curing the hide, and the labour of cutting and sewing it into jackets and pants, bartering for ready-mades was too attractive. They still killed animals—far more, as it turned out—but now traded the pelts for Hudson's Bay blankets and continental cloth. Animal furs became articles of commerce rather than seasonal gifts from the Great Manitou.

Today the clothing industry goes its profitable way almost independent of weather, relying instead on fashion gurus, catalogues, and on-line shopping.

For tourists wondering what to pack in the Maritimes, July and August can bring scorching 30°C plus temperatures—but sea winds cool most afternoons. Autumn offers the sweetest weather, with crisp sunny afternoons and frosty nights.

13

Weather

& Shelter

Around 1000 A.D. an Inuit hunter paddling his skin-covered kayak in Pistolet Bay saw a strange sight to seaward. A sleek brown thing with upturned neck and tail and a square wing rose over the eastern horizon and bore down on the low wooded coast like a great bird. Behind it appeared other such birds. The hunter paddled rapidly out of sight.

That summer a group of Vikings planted a settlement at L'Anse aux Meadows near Newfoundland's northern tip. It was a good time to be doing so, for the northern hemisphere was enjoying an interlude of milder weather, the same interlude that had helped their farming settlements in Greenland to flourish. The pack ice had retreated, making travel and fishing easier. Winters were shorter and milder, making it possible to start gardens and harvest food before winter.

Aboriginal Dwellings

For several years the Norse lived comfortably at L'Anse aux Meadows in their sod-covered timber-framed huts. Compared to the aboriginals' snug yet airy wigwams, these low dwellings must have been damp and smoky during wet weather. On the other hand, their rooftop covering of grass must have kept them cool in summer. The Vikings' abandonment of l'Anse aux Meadows within a decade was not due to inadequate housing, but to battles with the Inuit, who were harassing them with increasing intensity. Outnumbered, the Norse reluctantly sailed back to Greenland, leaving their camps to be discovered over nine hundred years later.

...its walls, which transmitted a very pleasant light, gave it an appearance far superior to a marble building...
—Arctic explorer Captain John Franklin, describing Inuit igloo interior, 1820.

PIONEER LOG CABIN (c.1700)
(NOTE HOME-MADE SHINGLES
& FIELD-STONE CHIMNEY)

AN ARCTIC IGLOO (SNOW)

Arctic explorer Captain John Franklin was so taken with the igloo that he remarked in 1820,

They are very comfortable buildings... one might survey it with feelings somewhat akin to those produced by the contemplation of a Grecian temple...both are triumphs of art, inimitable in their kinds.

A SUMMER WIGWAM
(BIRCH BARK & SPRUCE POLES)

The natives who drove them away housed their families in leather-shinned domes framed with whale ribs and driftwood. The northernmost Inuit summered in such huts but in winter they lived in elegant geodesic domes of snow made from precisely cut and fitted blocks of hard snow. So snug were these igloos that, with the warmth from the occupants' bodies, a single seal oil lamp sufficed to heat the interior.

Far to the south, the woodland Innu and Algonquians wintered just as snugly in two-layered, fur-lined conical birchbark wigwams with a central hearth, smokehole, and airflow-adjusting door. These houses could be put up in a few hours, taken down in less, and easily transported. The supporting poles were usually left in place for next time.

The extinct Beothuk of Newfoundland also summered in wigwams (mamateeks), but they wintered Athapaskan style, in family groups that shared lozenge-shaped birchbark-and-timber dwellings. These were sunk partly into the earth and heated by a hearth fire down the middle.

Settlers' Housing

Take a look at the mansions along Fredericton's Waterloo Row, or perhaps some homes in the north end of Halifax, or in the posher parts of Saint John or Charlottetown. While viewing their baroque ironwork gates, wide verandas, perky cupolas, well-trimmed lawns, and lush rhododendrons, it's difficult to imagine our ancestors living in log huts and drafty frame dwellings.

But they did. Doing so wasn't such a shock to them as we might imagine. In the seventeenth and eighteenth centuries the one-room family hut was still the norm for ordinary folk, and had been for generations. In fact those huts were often more comfortable than the dim, drafty, hard-to-heat, echoing castles where lords and ladies lived.

The peasant hut was a one-room kitchen-bakery-workshop-dormitory centred on the single chimney and a yawning fireplace. Except for some differences in materials, things were much the same for the peasant Acadians who settled Nova Scotia's Annapolis and Minas shores in the 1600s. Though their roofs were thatched with cordgrass from local salt marshes, the chimney still dominated one large room.

Because stone was scarce near the marshes, these chimneys were often built of short logs laid criss-cross in cabin style. To prevent the wood catching fire, it was plastered thickly inside and out with clay or gypsum. This plaster flue rose from a huge

ACADIAN HOME & BARN (c. 1670)
(NOTE THATCHED ROOF, ANIMAL SHELTER AT
RIGHT REAR, & ROOFED OUTDOOR OVEN)

fireplace that functioned as heat source, roasting spit, hot water heater, and clothes-drier. Connected to it on the outside wall was a roomy oven.

The chimneys of British and other settlers were built of stone or brick, but otherwise they fulfilled the same multiple uses. That gradually changed after 1740, when Benjamin Franklin invented his "Pennsylvania Furnace," the first free-standing indoor cast iron stove. Realizing that fireplaces sucked much of their heat up the chimney, he in effect shifted the fireplace out into the room. By enclosing the fire in a cast iron box and venting the smoke through a metal pipe, he got far more heat out of less wood. By putting the box on four legs, he generated a continuous circulation of cool air underneath and hot air overhead. By adding two doors, he retained the pleasure of watching an open fire.

ONE-STOREY
(WOODEN)

The Franklin Stove (as we now call it), when equipped with a coal grate, made it feasible to heat rooms separately without the mess and bother of having a fireplace in each. Thus stoves fostered the multi-room house—though several stoves sometimes consumed as much fuel as one big fireplace. The concept of the "master bedroom" developed soon after. For many years only the well-to-do could afford such luxuries, but by the mid-1800s the single-room dwelling was disappearing except on the frontiers and among the destitute. Ben's "furnace," essentially unchanged, is still with us.

TWO-STOREY (BRICK)

Roof Talk

Architecturally clever as we are, we sometimes neglect fundamentals that pioneers never forgot. At 11:00 A.M. February 3, 2001, about 150 parents were watching a Saturday junior hockey tournament between Tatamagouche and Amherst County teams in Springhill, Nova Scotia.

1960ˢ BUNGALOW

The Newfoundland
expression "tilt" refers to a
temporary lean-to, and
when they say someone
"smokes like a winter tilt,"
it's pretty serious. A lean-to
is normally sloped away
from the wind with the fire
facing in, so smoke in the
eyes is almost guaranteed
until the fire dies down.
Unless the wind changes—
and then it's cold.

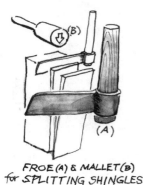

FROE (A) & MALLET (B)
for SPLITTING SHINGLES

Suddenly, with a crack like gunfire, a roof truss snapped. The crowd surged for the exits. Thirty players on skates beat them to it. As the last of them scooted out, a big section of the roof buckled under the weight of snow, and windows began to pop like corks, spraying shards of glass across the parking lot. Fortunately, no one was hurt.

Other east coast towns reported similar incidents in February. Flat-roofed schools and depots built in the 1960s were the worst. The same day the Springhill arena caved in, a government depot roof collapsed in St. John's, and a mall roof in Halifax sagged.

In case anyone had forgotten why the pitched (peaked) roof has been standard for northern homes for centuries, the winter of 2000-2001 was a sharp reminder. Suddenly rooftop snow removal became a priority. Hardware stores couldn't keep shovels and telescope-handled roof rakes in stock.

Steep roofs were customary in the snowy east for a very good reason—a half metre of snow and ice on a flattish roof is very heavy. The steep-pitched *habitant* roofs of old Quebec hark back to a time when thatching was common. Though Canada is rich in wood, settlers from wood-starved France continued to thatch for decades, likely out of habit.

Thatch leaks like a basket, so how can it keep a building dry? The trick was to slope the roof steeply enough for the water droplets to reach the eaves before falling through. The same principle applies to an emergency lean-to made of evergreen boughs. In fact, evergreen trees use the same pattern to shed snow. The steeper the slope, the thinner the covering need be— but the less headroom you have.

The lean-to is part of our culture. Nowadays we associate it with Boy Scouts and Girl Guides, but in pioneer days it was often the first shelter. The noted Canadian canoeist Bill Mason used the same design in his patented wilderness tent, adding an awning for outdoor cooking and eating.

Something that puzzled me when I first saw older homes in eastern Quebec was their lovely upturned pagoda eaves. I now think they were to prevent avalanching snow from ripping off the thatch and burying the windows.

Atlantic Canada's English settlers preferred to settle on the uplands where wood and stone were plentiful. For roofing they split wooden shingles or shakes from short bolts, using a froe (L-shaped iron tool) and mallet. To protect the walls they likewise adopted the steep-pitched, wide-eaved roof—except in Newfoundland, where the flattish roof became standard for all but the merchant class. Cost was probably the main reason, and they were easier to tar. Snow buildup wasn't a large problem because coastal gales do a good job of sweeping rooftops clear.

But you can't design a roof for winter only. There's the summer heat to consider. The best way to cool a house, short of surrounding it with shade trees, is to give it a tall attic. This provides an insulating dead air space to buffer the hot sun of July and August, which can be intense away from the sea's tempering breezes. The attic also provided safe dry storage of valuables.

A second line of defence was the high ceiling, which allowed overheated air to rise well above the occupants' heads. Such rooms were harder to heat in winter, but once heated they supplied a reservoir of warm air for the stove to keep in circulation. Little-used rooms were kept closed and unheated, since frozen water pipes were not yet a problem.

Another solution to summer heat was the outdoor kitchen. This was simply a cooking area, with its own chimney, separated from the main house by an airy passageway.

And of course our forebears knew the worth of shade trees. Newlyweds were reminded of this by the custom of planting a "His" and "Hers" broadleafed tree—sugar maple, ash, and oak were favourites—on either side of the entrance. Once grown, they helped shade the house from the morning and noonday sun. The also provided some protection from lightning, especially for houses built on a rise.

A house so equipped, with a couple of sugar maples or an elm to the south and east, was quite comfortable. I know, because we live in such a house. From 10 A.M. to 3 P.M., the hottest part of the day, our elm shades the whole roof.

Window Wisdom

Like roofs, windows say much about local weather. Pioneer Maritime homes had small windows and smaller panes, as glass was either unobtainable or unaffordable. At first people made do with lambskin (parchment) or oiled paper stretched over the openings. Parchment was durable but sagged in wet weather; paper tore if you looked at it sideways. Both let in light, but you couldn't see through them. Still, they kept out the wind and rain and snow, and no one could watch you bathing.

Window glass remained scarce and dear in the Maritimes for decades. It was costly because early glass was hand-poured or hand-blown. Shipping it overseas or carting it over rutted roads was chancy even for small panes. Your glass might arrive, but smashed to smithereens. Small panes were easier to replace and lost less heat. Big windows were a mark of wealth.

The day before central Nova Scotia maple syrup producer George Cooke was to start putting in his 25,000 taps for the 2001 season, his 6x12-metre (20x40 ft) sap storage shed collapsed under the weight of accumulated snow and ice. He had to sit out the season. But he replaced the roof, repaired his six stainless steel storage tanks and was back in business for 2002.

Newfoundland's earliest settlers were for several decades forbidden to install a permanent chimney or to put glass in their windows. This was to discourage year-round settlement and thus protect the lucrative Bristol merchants and crews, who came out when navigation opened each spring and returned home to England with their catch each fall.

As stoves proliferated and houses enlarged, windows became more important for light and for ventilation, especially in the summer. But because roughly 25 percent of a building's heat escapes through its windows, it became necessary to add storm windows for winter. In wealthier neighbourhoods shutters also became common, first to buffer winds and keep out burglars, and then for decoration.

Large or small, the traditional handmade wooden window is a work of art. There's the window box itself, an intricately-fitted wooden frame with a lower step sloped just enough to shed water. There are the slim dovetailed sashes, each one rabbeted (grooved) on the weather side to take the "lights" (as a *glazier*, one who puts glass in windows, would call the panes). The panes are fastened in the rabbets with glazier's points or *sprigs*, and sealed inside and out with neatly trowelled waterproof putty, a dough-like mixture of chalk and linseed oil that hardens.

There are various window models: single-hung (one-piece), double-hung (with a sash that slides up for ventilation), windows hinged on either side (casement style), hinged above or below, and so on. Oldtime double-hung windows came with ingenious swivel locks of decorated cast iron or brass that fastened the lower sash to the upper. Some had side stops to hold the lower sash at any level, matching storm windows that were taken off each spring, and insect screens for summer.

The wood used for all these interlocking pieces was white pine, the lightest, softest, most workable of native woods, and the one least likely to rot, warp or split with time. There are pioneer window frames that still work perfectly after three centuries.

Even with small windows, heating an ordinary house required twelve to fifteen cords of firewood or several tonnes of coal a year. Naturally, homeowners tried to insulate. Lacking fibreglass, they devised other solutions.

We discovered this when we bought our 1850s farmhouse. To provide a sealed air space between the outer sheathing and the plastered wall, the builders had nailed laths across the vertical studs on the *outside* and plastered them, a process called rendering. Combined with the inside plaster, this created a 15cm/6in layer of dead air. This method insulates far better than bare boards, and helps cool the walls in summer.

A somewhat less elegant solution was to stuff the walls with buckwheat hulls. Early grist mills ground a lot of buckwheat (*Fagopyrum esculentum*), a relative of rhubarb which settlers

Very old windowpanes are thicker at the base. This is because the glass behaves like very cold molasses, moving downward as the molecular layers slide over one another.

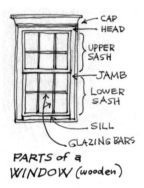

CAP
HEAD
UPPER SASH
JAMB
LOWER SASH
SILL
GLAZING BARS

PARTS of a
WINDOW (wooden)

The word window comes from Middle English *windoge* via Old Norse *vindauga*, "wind eye."

found useful not only for making pancake flour, but as a green manure crop for breaking down sod and shading out weeds in newly cleared fields. The beech-nut shaped hulls were durable, air-filled, unattractive to vermin, and easy to pour into small spaces. Moreover, they were free for the taking at local grist mills. Builders carted loads of them to the work site, filled the spaces between the studs with them, and sealed them in with hemlock or spruce planking under clapboard or shingles.

Another excellent insulating material used on both sides of Northumberland Strait was eel-grass (*Zostera marina*). This pliant sea plant thrives in shallow, sheltered coves with silty bottoms, and is a favourite food of Canada geese, whose rich guano fertilizes the beds. Dead eel-grass, also free for the taking, washes up in long windrows along the beaches. Carted home and dried, it becomes springy, excelsior-like, and can be stuffed into or packed against house walls. Since eel-grass is saturated with iodine and salt, it doesn't rot the wood.

A third material used for insulation was dry sawdust. This was definitely a third choice, because over time it tended to clump toward the bottom, leaving the upper spaces empty. But where buckwheat or eel-grass were scarce, sawdust was acceptable. It smelled nice, too, and once under cover didn't rot.

The house I grew up in had sawdust in its walls. A good thing too, for our house had been moved from elsewhere, which opened the seams and made it drafty. My grandfather operated a sawmill across the road, so my father filled the walls with it. Over time the sawdust settled, and some sifted out whenever a door was slammed. I used to thump the clapboard to see it fall like snow—and usually got scolded for my mischief.

The Planters brought several other domestic innovations. To solve the problem of wading or shovelling through snowdrifts to do barn chores, they linked house, barn, and outbuildings via roofed passageways. In dirty weather, householders could feed and milk the cows and tend the livestock without ever stepping outside. These corridors had doors to seal drafts and odours out. Such a dwelling was almost as weather-free as a shopping mall.

JOINED (FARM) BUILDINGS

Linked structures weren't entirely a New World innovation, however. In the Middle Ages, Britons and Europeans commonly stabled livestock, including cows and pigs, in their homes, with only a hallway between. This system later evolved into a sort of semi-circular compound around a courtyard, with wide eaves to keep out the weather. New Englanders, faced with a harsher climate, extended this to its logical conclusion. It was a lot more sanitary.

The Saltbox

When I was a boy there was an old, grey house standing on an island near the mouth of the river. The story was that this was the first house built in our bay, that it belonged to a family of salmon fishers who wanted a measure of protection from the native Beothuk. The natives understandably saw them as poachers and harassed them whenever they got the chance. Indeed, the little cemetery had a stone marking the final resting place of a fisher they had beheaded.

What I didn't know then was that this house was an authentic saltbox built in the early 1800s, a fact confirmed by the newspapers under the wallpaper. The saltbox design was developed in New England in the seventeenth and eighteenth centuries and owes its name to those roadside boxes used for storing salt and sand. Two full stories high on one side and only one on the other, it featured a long roof that made it look lopsided.

The long roof not only deflected winter winds, but provided a combined storage and porch area much larger than the conventional add-on porch. Usually the wider slope faced north or northwest. Sometimes, for even greater protection, the house was built against a bank or cliff. Such a dwelling, with its walls sheathed in birch bark and filled with seaweed or sawdust, was fairly snug, especially after a blanket of snow built up on the roof. According to historian Lewis Mumford, it equalled in comfort any British homestead of that period.

SALT BOX (WOODEN)

The long lean-to porch not only kept snow and drafts and mud out of the house proper, but provided, in those days before electric freezers and indoor plumbing, ample and accessible storage. As with the New England connected house, it allowed one to fetch fuel, water or produce without having to go outside. It served as a pantry for nonperishables like summer herbs, flour, lentils, dried fruits, and smelly smoked meats and fish. During the coldest part of the winter, it stored haunches of fresh beef, pork, or venison, which kept rock-hard well into March. But the water barrel, being in constant use, merely iced over.

Covered Bridges

One of most amazing sights of my first year in the Maritimes was the covered bridge. It so happened there was one near where my girlfriend lived in eastern New Brunswick. It was over 100m/300ft long and spanned the West Branch of the St. Nicholas River where it enters the Richibucto.

At first glance the structure looked improbable: a very long grey shed built over the water? Why? I assumed it was to shelter horses and people crossing by sleigh or on foot. So it did; but the main reason was to keep the snow from blowing away or melting on the roadway. Horses couldn't pull loaded sleds over bare wood. Some covered bridges had small pane-less windows that let in both light and snow, but most relied on loosely spaced vertical planks, while the roof kept out rain and sun. Such a bridge might stay well iced from November through April. It created a winter microclimate.

Stern and puritanical our ancestors may have been, scornful of novelties that served no purpose beyond embellishment, yet they were quick to seize on improvements to their living conditions. In part, this open-mindedness came from having left the lord-and-serf system overseas. Each family could now own at least a few acres and enjoy access to common land. Also, within the bounds of local bylaws, they were free to change their structures to suit the weather and the times. What was good they kept, and what they kept they improved on, until the New England village was a model of practical beauty and architectural efficiency.

These weather-related improvements were exported into eastern Canada not only by the Acadians and Planters, but by the flood of Loyalists following the American Revolution. Conjoined farmsteads can still be found from Ontario east, and a few covered bridges still grace the New Brunswick countryside.

LOYALIST SQUARED-TIMBER HOME (c.1860)
(NOTE BRICK CHIMNEY & DUG WELL)

The Modern House

We have come a long way with our nesting instinct. In a way the modern detached house is a testament to our success in coping with a demanding climate. Each part attests to this. The house sits on cement footings dug below frost level to prevent heaving. Outside the footings are tubes of vinyl or ceramic to drain groundwater away from the basement. The innards are insulated from basement to attic; windows are airtight, triple-glazed and equipped with insulated drapes (with the Californian picture window facing south). There is central heating and, increasingly, in-floor electric heating, electric heat pumps, and/or solar panels on the roof. The cedar, asphalt or metal ("Lifetime Guarantee!") roof shingles have ice shields to prevent meltwater backflow; the vinyl or metal eavestroughs and downspouts carry cloudbursts clear of the walls. Grounding wires prevent lightning damage; the vinyl siding is weatherproof ("No more peeling paint!"); airtight, super-efficient stoves provide supplemen-

Would it be an exaggeration to say that there has never been a more complete and intelligent partnership between the earth and man than existed, for a little while, in the old New England village?

—Lewis Mumford, *Sticks and Stones: A Study of American Architecture and Civilization*

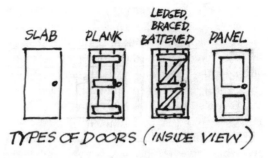

SLAB PLANK LEDGED, BRACED, BATTENED PANEL

TYPES OF DOORS (INSIDE VIEW)

tary heat, supplanting the mostly decorative fireplace. If anything testifies to our unrelenting contest with climate, the house is it. Of course we know that a fierce flood can easily wash it away, that a powerful tornado can smash it to splinters. But, like the moose with two-metre-wide antlers running through trees a metre apart, we do the best we can.

Weather Wear & Tear

Weather makes other inroads into our peace of mind. In Alistair MacLeod's 1999 novel *No Great Mischief*, the narrator's sister visits an old relative in her stone house in the Highlands. The old lady wonders how Canadians can live in wooden houses. "Don't they rot?" she asks.

They do, of course. Year in, year out, weather nibbles away at shingles and paint and, yes, even vinyl and metal. It does so by weathering, whereby tiny particles are lifted by the steady scuffing and rubbing of wind and snow and rain, and by daytime swelling and nighttime shrinkage of paint and wood through heat and cold. The paint, already thinned by weathering, gradually crackles and peels, exposing the wood.

The real culprits are sun and wet. Wood actually keeps longer without paint, but then weathering strips the surface fibre by fibre, unhindered. Ten years of this, and the grey boards, though lovely, resemble an old school floor with all the nails and knots in bas relief.

Before the advent of long-lasting siding, builders clad almost all frame buildings with spruce clapboard. Each ship-lapped board protected the one beneath, and a dripboard protected the baseboard. They then primed the wood with flat white paint and finished it with a coat or two of lead-based exterior paint glossy enough to resist weathering, first a thin coat, then a thicker one. It wasn't a bad system. It's still used by those who can't afford siding or don't like to wrap their homes in plastic. Vinyl is expensive to buy, but lasts for more then twenty years. Then, having been exposed to heat and cold and ultraviolet for too long, it turns brittle. Still, many homeowners swear by it, and metal is even more durable.

On our frame house, my paint jobs last from two to five years. The north wall lasts longest because the sunlight never reaches there. The next in durability is the east wall, because the sun only shines there for a few morning hours at a time. The worst are the south and west walls, which get the most sun; they start to blister in two years or less. One reason oldtime houses had so many trees about was to slow this weathering process by screening out direct sunlight.

Other homeowning headaches can also be traced to weather. Rain gets under flagstones, freezes, expands, and shoves them out of line and level. It loosens fence posts and stone walls, lifting and finally toppling them. "Something there is that doesn't love a wall," mused the poet Robert Frost—though he admitted that good fences made good neighbours. Our salty air, abetted by sulphur dioxide and other pollutants, loosens bricks and building stones by eating away the lime in mortar.

I used to wonder why old fence posts on sidehills always seemed to lean downhill. The reason, I later learned, is *solifluc-tion*—the alternate expansion and contraction of the earth due to solar heating and nighttime freezing. Water expands when it freezes. The bottom of a fence post is pointed like a pencil, so each time the earth expands, it shoulders the point upward a few millimetres. The same process pushes rocks up into farmers' fields, and lifts container seedlings out of the soil. On a south-facing slope with lots of sun, fences have to be set upright every few years.

Weather also plays havoc with our lanes and roads. Only by continual grading, regular filling of potholes, and yearly mending of frost cracks can they be kept driveable. A paved road or an airport runway left untended for five years will develop many cracks around the edges from frost heaving, and small potholes elsewhere. Leave it another five years, and lichens, moss, and grasses will have colonized the cracks. Another five years, and woody plants like goldenrod and alder will be taking hold. After that the trees will come, first light-lovers like aspen, grey birch and larch, then shade-tolerant species like maple, yellow birch, and fir.

It is as if the weather and plants were in cahoots.

14

Weather
in Words

In the summer of 1816, "The Year Without a Summer," the young English poet Percy Shelley and Mary Shelley were vacationing with a group of friends on Lake Geneva in Switzerland. Disgusted by the cold, rainy weather that kept them indoors, they started telling ghost stories by the fire. This led to a contest to see who could write the best tale of horror. All agreed that Mary's was the best. She called it *Frankenstein*. Published two years later, it became one of the world's favourite horror stories.

The real monster that summer was the Indonesian volcano Tamboro. Erupting 14 months before, it polluted the atmosphere with dust and aerosols that veiled the sun and brought on a sort of nuclear winter. Had Switzerland's summer been normal that year, *Frankenstein* might never have been written.

Weather colours our culture and language more than we perhaps realize. How often do we hear expressions like these?

He ran like the wind.
The boss is in a fog today.
Parliament has just endured a stormy session.
Well, they got a cool reception.
The conductor received thunderous applause.
She was dewy-eyed with sentiment.
Oh, that's lost in the mists of time.
Do you think he'll weather the storm?

The grey and slanting rain squalls have swept in from the sea and then departed with all the suddenness of surprise marauders.

—Alistair MacLeod, *The Lost Salt Gift of Blood*, 1976

Rarely a day goes by that we don't hear some undecided person say "I'll take a rain check on that." Irked by criticism, we say, "Don't rain on my parade!" Someone who deserts us at the first sign of trouble, we call "fair-weather friend." Defiantly we shout, "That'll be a foggy Friday!" Suddenly in love, we're on "Cloud Nine."

Weather lives in the news, on the street, and in pop culture. We meet it in the titles of movies (*The Perfect Storm, Hurricane*), and of books (*Isaac's Storm, Noah's Flood*). And of course in music.

Weather in Music

Music is perhaps our richest repository of weather titles: "Stormy Weather," "You Are My Sunshine," "Clouds," and "Rainy Night in Georgia." The venerable "Somewhere, Over the Rainbow" owes much of its charm to the ancient Celtic belief in the leprechaun, a troublesome mini-human said to bury his crock of gold ransom money under a rainbow. In his "Subterranean Homesick Blues," streetwise balladeer Bob Dylan took a common phrase and used it to caution the unwary: "You don't need a weatherman to know which way the wind blows." Joni Mitchell made weather a metaphor for "love's illusions" in her 1969 hit album "Clouds," rereleased under the same title in 1999. Nova Scotia singer-songwriter Anne Murray transformed the intrepid little snowbird into a symbol for love's stormy weather.

The composers of hymns are notably fond of weather metaphors—especially storm metaphors. "Will Your Anchor Hold in the Storms of Life" and "From Every Stormy Wind that Blows" come to mind. On a gentler note we have the 1760 composition "Lo, He Comes with Clouds Descending," and Jessie Adams's 1908 classic, "I Feel the Winds of God Today." One of the Western world's all-time favourite carols, Hans Gruber's "Silent Night, Holy Night," has inspired millions of sappy Christmas cards. Recently, there's the lovely "Huron Carol," which begins, "`Twas in the moon of wintertime."

Weather in Everyday Speech

Weather has woven itself into the very nouns, verbs, adjectives, and adverbs which are the armature of language. From the Second World War and the Battle of Britain we got the *Hurricane*, a fast, propeller-driven fighter plane. The world of salmon angling has an artificial fly called "Thunder & Lightning." And look how the noun "snowbird" has spread its wings. My dic-

Bing Crosby's 1942 version of "White Christmas," written by Irving Berlin for the movie *Holiday Inn*, is the best-selling record in history, with over 35 million copies sold. The movie was reissued in 1954 as *White Christmas*.

Quebecers call the snowbird "oiseau de misère," meaning "bad weather bird."

tionary defines it as the snow bunting, a migratory boreal finch which winters along our coasts and comes inland during early blizzards. The word also denotes a species of well-off Canadians who escape our winters by migrating to Florida for a few months.

The world of publishing has a tool called the "fog index" which they use to rank the clarity of prose. Bad prose typically relies on long sentences, big words and convoluted sentence structure. By assigning each a numerical value, the index provides a measure of how easy a given piece of writing is to decipher. The worse the prose, the higher the score. Thus, a low fog index writer might say, "We found signs of wrongdoing," while a high index writer might say, "Upon investigation it was ascertained that meretricious behaviour had transpired"—or worse.

People often use the nouns "atmosphere" and "climate" to describe human emotions and behaviour. When tensions run high, we say the atmosphere is "electric" or "overcharged." Newscasters tell us that "the atmosphere in the hostage-taking drama is tense this morning." (Though atmospheres never get tense.) Or we're told that "President George W. Bush took office in a climate of dissatisfaction." We say that an athlete has "lightning reflexes." The expressions "snowy owl," "snow goose," "snowbound," "frosty stare," and "chilly reception" owe much to weather.

Rural folklore is typically loaded with weather references. "Farmer's fertilizer" refers to a late spring snowfall that soon melts. I used to interpret this to mean that the snow watered the ground, supplying nitrogen. The likelier meaning is that it insulates the newly thawed ground against further freezing, thus speeding the growth of grass and fall-seeded crops. Sheep farmers call an unseasonably cold May wind a "lamb killer." Forest managers say "windthrow" and "blowdown" for trees toppled or broken by gales.

Place names often carry weather connotations. The East Coast has at least two Blomidons, the best known being the promontory that dominates Nova Scotia's Minas Basin between Wolfville and Parrsboro. Its name comes from "blow-me-down," sailor talk for a notoriously windy place. In the days of sail, Blomidon's lethal combination of fierce tide rips and sudden downdrafts capsized many a ship.

Wreckhouse Cove on Newfoundland's southwestern corner has a similar reputation. Local winds of up to 140kmh/84mph, funnelling off nearby Table Mountain, have flung boxcars off the railway tracks and tumbled tractor-trailers off the Trans

Canada Highway. For years the Newfoundland Railway (and later Canadian National) paid Lauchie McDougall of Wreckhouse up to $140 a year—one dollar per kmh?—to say whether it was safe to pass. McDougall did this for 30 years until he died in 1965. One time CN ignored his advice; 22 boxcars blew off the tracks.

If you ask Acadian fishers of western Cape Breton Island what wind they fear, they will tell you about the *suete*. The name means southeaster (from *sud est*), a wind so fierce in those parts that fishers routinely double its forecasted velocity. It develops when a westerly warm front blocks easterlies coming off the highlands, causing them to spurt sideways like water from a partly blocked garden hose. This wind has drowned fishermen, shifted houses off foundations and torn slabs of roof off buildings. A notably fierce one recently demolished an Irving storage tank, forcing the residents to leave until the mess was cleaned up.

Lauchie McDougall of Wreckhouse on Newfoundland's southwest corner could "smell" high winds coming. His nose was so reliable that CN paid him to advise whether it was safe to send the train along that stretch.

The *suete* off Cape Breton Island's west coast has been clocked at 194kmh/120mph.

Weather in Literature

Sacred and profane literature in all languages is full of weather. The Hebrew and Christian bibles are notably rich in it. The Noah flood narrative comes to mind, with its rainbow signalling God's promise (testament) not to drown the earth again. In the First Book of Kings there is the aforementioned dramatic story of Elijah and the priests of Baal. There are some splendid storms: the one with Jonah and the great fish, the

Galilee storm which Jesus calmed, and the one which the apostle Paul barely survived on his journey to Rome.

Because the Israelites were a pastoral people, weather was often used in blessing or cursing. The dying Moses, saying farewell as they enter Palestine, says: "May the Lord open the heavens for you, his rich treasure house, to give you rain upon our land at the proper time and bless everything to which you turn your hand" (Deut. 28:12). Should they forsake the Lord God, he left them with this malediction: "May the skies above you be bronze, and the earth beneath you iron. May the Lord turn the rain upon your country into fine sand, and may dust come down upon you from the sky until you are blotted out" (Deut. 28: 23-4).

Of the Bible's 66 books, probably the richest in weather allusions occur in Job, Psalms, Proverbs, and Ecclesiastes. In the Book of Job, written as early as the tenth century B.C., God asks a rhetorical question:

> Has the rain a father?
> Who sired the drops of dew?
> Whose womb gave birth to the ice,
> and who was the mother of the frost from heaven, which lays
> a stony cover over the waters...?
> —Job 38: 28-30, NEB

There is even an early reference to the hydrological cycle:

> He draws up drops of water from the sea and distils rain from the
> mist he has made; the rain-clouds pour down in torrents....
> —Job 36: 27-28, NEB

The 11th chapter of Ecclesiastes warns us that "He who watches the wind will never sow, and he who keeps an eye on the clouds will never reap" (v. 4, 5). In Solomon's Song of Songs (2:11-13) there is this lovely passage:

> For now the winter is past,
> the rains are over and gone;
> the flowers appear in the country-side
> the time is coming when the birds will sing,
> and the turtle-dove's cooing will be heard in our land....

The late Canadian literary critic Northrup Frye argued forcibly in both *The Great Code* and *Words With Power* that the Bible has coloured all of Western culture, for believers and unbelievers alike. It has certainly elevated our early perceptions of weather to almost the supernatural, as have ancient Greek legends. Homer's *Iliad* and *Odyssey* (circa 700 B.C.) accord the gods such control that Mount Olympus, their home,

> ...is neither shaken by winds nor ever wet with rain, nor does snow
> come near it, but clear weather spreads cloudless about it....
> —Homer, *The Odyssey*, c. 700 B.C.

Nourished by such sources, European literature is a trove of weather prose. From thirteenth-century Italy we have this lovely canticle from Francis of Assisi, a walker in all weathers:

> Praised by the Lord
> For our Brother the Wind,
> And for Air and Cloud,
> Calms and all Weather.

One reason why the works of Danish novelist Knut Hamsun (1859-1952) give me such pleasure is his passionate treatment of the passing seasons, especially in his *Growth of the Soil*. I feel the same passion in the work of France's Jean Giono, and in Vladimir Nabokov's memoir of his Russian childhood, *Speak, Memory*. For me, the weather in Boris Pasternak's *Doctor Zhivago*, as in the novels of Dostoevsky, feels comfortably Canadian. Perhaps it was for the same reason that I didn't enjoy Thomas Mann's *The Magic Mountain* until I got to his marvellous mountain blizzards.

British literature, especially Romantic literature, has shaped our weather attitudes more than any other. This is true even if we haven't read the works, for they permeate our culture as the weather permeates the works. Good novelists, said Hemingway, never leave the weather out.

What would Shakespeare's "The Tempest" be without the storm? Thomas Hardy's splendid heroine Eustacia of the cloudy Wiltshire moors in *Return of the Native* wouldn't be half as bewitching without the masterfully painted weather backcloth. Likewise Dickens's *David Copperfield* needs that raging coastal storm near the end to sweep away the accumulated nastiness.

Few Britons have written so eloquently of weather, especially the English spring, as John C. Powys (1872-1963). In his novel *Wolf Solent* (1929) one can fairly smell and feel the rain and mud.

When our children were young, we delighted in reading to them from A.A. Milne's *Winnie the Pooh* series, especially this verse from *Now We Are Six*:

If I were a bear
  And a big bear too,
I wouldn't much care
  If it froze or snew.
I shouldn't much mind
  If it snowed or friz
I'd be all fur-lined
  With a coat like his!

I also tried to read them the Welshman Dylan Thomas's *A Child's Christmas in Wales*, because he immortalized snow in childhood—but they were too young for its linguistic gymnastics. Now the CBC broadcasts this prose poem every December with East Coast actor Gordon Pinsent narrating.

Polish novelist Jozef Korzeniowski, alias Joseph Conrad, waxed poetic over marine weather. In the late 1800s he wrote some magazine sketches which later became his autobiographical *Mirror of the Sea*. In it he portrays the west wind as a god:

Clothed in a mantle of dazzling gold or draped in rags of black clouds like a beggar, the might of the Westerly Wind sits enthroned upon the western horizon with the whole north Atlantic as a footstool for his feet and the first twinkling stars making a diadem for his brow. Then the seamen, attentive courtiers of the weather, think of regulating the conduct of their ships by the mood of the master.

Compare that to A.E. Houseman's saccharine "It's a warm wind, the west wind, full of birds' cries/I never hear the west wind, but tears are in my eyes...", or even to Keats' "Ode to the West Wind," and we know who is the true weather poet.

English poetry is if anything even more weatherful than English prose. Geoffrey Chaucer set the tone in "The House of Fame" (1374-1385) with lines like:

Lord, this is a huge rayn!
This is a weder for to slepen in....

No one has ever portrayed the seasons as truly as William Shakespeare. His line "When icicles hang by the wall...and milk comes frozen home in pail..." (Sonnet II) is winter personified; his "Where the bee sucks, there suck I/In the cowslips bell I lie..." (Sonnet IV) is fragrant with summer.

And John Bunyan, in his vastly popular *The Pilgrim's Progress* (1678), expressed his devotion to truth in terms of weather:

One here will constant be,
Come wind, come weather.

Many later poets, notably Burns, Wordsworth and Tennyson, have used weather to express love or lack of it. A generation of school children memorized Alfred Tennyson's tear-jerking "Lucy Gray." Less well known was Thomas Hardy's love note:

I need not go
Through sleet and snow
To where she waits for me;
She will tarry me there,
Till I find it fair,
And have time to spare
From company.

Robbie Burns is widely quoted for his love lyrics, but here's a verse one seldom hears, from "On the Birth of a Posthumous Child":

May He who gives the rain to pour,
And wings the blast to blaw,
Protect thee frae the driving show'r
The bitter frost and snaw!

I fancied William B. Yeats was rich in weather until I read more carefully; he isn't. But Samuel Coleridge in "The Rime of the Ancient Mariner" makes up for it in image after striking image:

And the coming wind did roar more loud,
And the sails did sigh like sedge;
And the rain poured down from one black cloud;
The Moon was at its edge.

The priest-poet Gerard Manley Hopkins (1844-1889), originator of what he called sprung rhythm, had this to say of the very air we breathe:

Wild air, world-mothering air,
Nestling me everywhere,
That each eyelash or hair
Girdles; goes home betwixt
The fleeciest, frailest-flixed
Snowflake; that's fairly mixed
With, riddles, and is rife
In every least thing's life....

Dylan Thomas's "Poem in October," written in 1944 on his thirtieth birthday, makes rain a metaphor for time:

And I rose
In rainy autumn
And walked abroad in a shower of all my days

and, further on,

A springful of larks in a rolling
Cloud and the roadside bushes brimming with whistling
Blackbirds and the sun of
October summery....

All of this has filtered into the consciousness of Canadian writers. Sinclair Ross's sinister blizzard in "The Painted Door" rings true for any northerner. First he gives us catspaws of fine snow swirling delicately across the roads, then the northeaster flexing its muscles, then the full three-day onslaught, and finally the northwesterly gale that leaves its drifted signature on every highway cut, barn, house, shed, and clump of grass. Probably the best Canadian chronicler of that post-blizzard world is Frederick Grove. His *Over Prairie Trails* is a classic.

Closer to home, Nova Scotia's Ernest Buckler uses snow to describe the onset of death in the lovely closing scene of *The Mountain and the Valley* (1952):

> *The snowflakes fell on David's face and caught in his eyelashes and melted....And then they did not melt on his eyelids or on his cheeks or in the corners of his mouth or anywhere on his face at all.*

And we have Hugh MacLennan describing Halifax the morning after the great explosion of December 6, 1917, in his 1941 novel *Barometer Rising* (note the title):

> *The streets were deep in drifts and gashed darkly by the marks of horses, sleighs, and human feet, and in the downtown districts people shuffled aimlessly through uncleared sidewalks past boarded and half-empty stores, their overcoats appearing so black against the dreary whiteness of the streets and the sooty red bricks of the battered buildings that they all seemed to be in mourning.*

And Thomas H. Raddall writes in *Hangman's Beach*,

> *...there were storms for days and nights on end, when they could do nothing but huddle with the farm folk by the hearth, hearing the boom of the wind in the chimney and the slash of snow on the panes.*

Nova Scotia novelist and poet Charles Bruce, father of journalist Harry Bruce, had this to say (in *The Channel Shore*) of late winter in Nova Scotia:

> *The same weather, cold and windy in March, bright with sun under tall clouds or overcast with brooding rain, from June to fall; frosty or wet or white with snow in winter. In all these usual things there was something to interest a man.*

Lucy M. Montgomery's weather, like the Prince Edward Island landscape itself, tended to gentility:

> *It was a long drive, but Anne and Diana enjoyed every minute of it. It was delightful to rattle along over the moist roads in the early red sunlight that was creeping across the shorn harvest*

*fields. The air was fresh and crisp, and little smoke-blue mists
curled through the valleys and floated off from the hills.*
—*Anne of Green Gables*, 1908

Newfoundland's Margaret Duley pivots her fiction on people
and place, yet she can evoke weather too:
*Summer died in a day. The gales tortured the boats and drew
angry crests on the waves. The trees shivered in the wind, and
when an early frost curled up their leaves they rustled with a dry
sound.*
—*The Eye of the Gull* (1976 reprint)

In our own time, New Brunswick's Antonine Maillet and David
Adams Richards wrap us in familiar weather. Maillet, descended
from exiled Acadians who returned from America to settle Kent
County's Buctouche shore, captures the mingled confusion and
exaltation of her people as they near the end of their thousand-
mile trek:
*Where did Maine end and Acadie begin? ... None of these home-
bound exiles could have told you by what channel they were
entering the country or when, precisely, they had crossed the border. Between the stormy autumn sea, the brilliant forests with
their complete palette from yellows to flaming reds, between the
warm, singing inland winds and the smell and rustle of dead
leaves beneath their feet, somewhere in the midst of all that,
Acadie lay concealed....*
—*Pelagie*, 1979 (tr. 1982)

From New Brunswick's Miramichi region comes the distinctive
voice of the early David Adams Richards:
*The streets were dark and windy, the rain lashing against their
faces. He could see the others shutting their eyes because of it,
looking as if they were in some terrible pain.*
—*The Coming of Winter*, 1974

Nova Scotia's Susan Kerslake evokes something of Dylan
Thomas's whimsy:
*Then into the night came the sound of the wind, beginning like
a gasp released out of a pin in the ground, from under a wave or
through a hole in the ink-black sky.*
—*Middlewatch*, 1976

Of late our feisty weather has been showing up in the work of
younger writers, sometimes in the guise of humour;
Newfoundland's Wayne Johnston is adept at this:

And he vowed that, as long as he was weatherman, the sun would never shine in Russia. He would rain down hail on Idi Amin. He would cause a cloud of sulfide smog to settle on Toronto, and would leave it there until all the Newfoundlanders got fed up and came home. He would be the first-ever cathartic weatherman.
—*The Story of Bobby O'Malley,* 1985

In the short stories of Alistair MacLeod, weather is not so much described as felt. Even in a thick novel like *No Great Mischief*, where the early Hugh MacLennan might have devoted half a page to describing a cloud, MacLeod gives us the briefest of descriptions. Yet the weather is there, pressing against his fictional kitchen windows, enfolding his characters. So accustomed are his homesick Cape Breton miners to checking the morning skies of home, that, even working underground in the wilds of northern Ontario, they still listen to the daily weather forecast. So when MacLeod actually does describe weather, it is doubly telling:

The grey and slanting rain squalls have swept in from the sea and then departed with all the suddenness of surprise marauders. Everything before them and beneath them has been rapidly, briefly and thoroughly drenched and now the clear droplets catch and hold the sun's infusion in a myriad of rainbow colours.
—*The Lost Salt Gift of Blood,* 1976

That grand old man of early Canadian poetry, Newfoundland-born Edwin J. Pratt, was big on weather. Of all his poems long and short, perhaps the best known is "Erosion":

It took the sea a thousand years
A thousand years to trace
The granite features of this cliff,
In crag and scarp and base.

It took the sea an hour one night,
An hour of storm to place
The sculpture of these granite seams
Upon a woman's face.
   —*The Collected Poems of E.J. Pratt, 1958*

More recent Canadian poetry abounds in weather metaphors. Thumbing through the landmark *15 Canadian Poets Plus 5* (1978), I was struck by the amount of snow and ice. The young Margaret Atwood made much of our northernness:

Why do you smile? Can't you
see the apple blossoms falling around
you, snow, sun, snow, listen, the tree
dries and is being burnt...

And Dorothy Livesay's "Summer Landscape: Jasper" starts out with:

The mountains are cold husbands
they stare
stony and white-capped
shrouded in mist

Al Purdy's "Detail" ends with:

one week in late January
when wind blew down the sun
and earth shook like a cold room
no one could live in
with zero weather
soundless golden bells
alone in the storm

Leonard Cohen, in "Disguises," bidding goodbye to some of his friends:

I loved your puns about snow
even if they lasted the full seven-month
Montreal winter. Go write your memoirs
for the Psychedelic Review

And Phyllis Webb couches a critique of the poet Rilke in these words:

> ...*I cannot take so much tenderness, tenderness, snow falling like*
> *lace*
> *over your eyes year after year as the poems*
> *receded, roses, the roses, sinking in snow*
> *in the distant mountains.*

Margaret Avison remarks,

> *The sun has not absorbed*
> *this icy day, and this day's industry—in*
> *behind glass—hasn't the blue and gold, cold*
> *outside.*

And here's Gwendolyn MacEwen, puzzling over the state of sleep:

> *Your breathing is a thing I cannot enter*
> *Like a season more remote than winter*

Patricia Lowther closes "Early Winters" with

> *Deer on the winter road*
> *wore jewels on their antlers*
> *the spines of the burnt mountain*
> *sifted the snow*
> *like a giant comb*

Virtually the only easterner in this pioneering anthology was the late Alden Nowlan, poet-in-residence at the University of New Brunswick. Oddly, his best snow poem was not about Canada at all, at least not on the surface:

> *Wind in a rocky country and the harvest*
> *meagre, the sparrows eaten, all the cattle*
> *gone with the ragged troopers, winter coming....*
> —"For Nicholas of All the Russias"

But for genuine Canadian wintriness, give me Nowlan's lines from a later book, describing Saint John, where he worked as a cub reporter on the *Telegraph Journal*:

> *It's snowing hard enough that the taxis aren't running.*
> *I'm walking home, my night's work finished,*
> *long after midnight, with the whole city to myself...*
> —The Best of Alden Nowlan, 1993

Actually, the East Coast is wetter than most of Canada. So our poems are wetter too. As Labrador poet Boyd Warren Chubbs writes,

Snow holds the stains
until the rains take them, bend
seasons to hide the day, draining the tough evidence,
the earth opening and closing...
—And You, Blessed Healer, 1996

Maritimers do rejoice when April washes winter's snow away. As Fredericton's young Robert Gibbs put it,

Rain breaks the snow
and streams off the steam
—A Kind of Wakefulness [Undated]

I'll let Harry Thurston have the last and wettest word:

O I like the world
like it is now
wet like after making love
rain brimming the barrel
dancing high off the tin roof
swelling all the cells
of all the leaves
—Clouds Flying Before the Eye, 1895

East Coasters can relate to that.

15

Weather

Lore &

Superstition

Stop a Maritimer young or old in a mall in downtown Charlottetown or Saint John or Halifax, and ask them to quote one weather saying. Ten to one, they'll come out with:

Red sky in the morning,
Sailors take warning.
Red sky at night,
Sailors' delight.

Or some variation of it. The reason so many people know this weather maxim is that it's one of the world's oldest. The ancient Greeks knew it. Jesus used it in a parable. And the reason it's still around is that it works—which is more than one can say for most weather ditties. For example, this one:

Saint Swithin's Day an ye do rain,
For forty days it will remain;
Saint Swithin's Day an ye be fair,
For forty days 'twill rain nae mair

I went out to see if there were any signs of my destiny in the sky, but there weren't—there was nothing but snowflakes.

—The Hon. Pierre Elliott Trudeau, describing his Feb. 28, 1984 walk in an Ottawa snowstorm to decide whether to quit politics. He did, that June.

Saint Swithin was a ninth-century bishop of England's Winchester Cathedral who made a dying wish not to be buried inside the church but out under "the sweet rain of heaven." His wish was granted, but on July 15, 971 A.D., some monks removed his body to a crypt in the new basilica. Forthwith a forty-day rain set in. July 15 has been called Saint Swithin's day ever since, making the saint a sort of summer groundhog.

But that was England. Surely this monk's curse hasn't tracked us across the Atlantic? I recall a summer in the 1980s when we *did* suffer five to six weeks of rain and fog—but at the time I didn't check the dates. The fact is, many of our weather axioms are based on British weather and are next to useless here. Another good example:

The North Wind doth blow
And we shall have snow.

Most Maritime snow comes from the east. And another example, regarding the month of March:

In like a lamb,
Out like a lion.
In like a lion,
Out like a lamb.

They're talking about springtime in Britain. No matter how March comes in over here, lamb or lion, we won't get real spring weather for another month, if then.

I don't blame our ancestors from Europe and Britain for importing these maxims. Old habits die hard. By the time they realized Maritime weather was different, the sayings had taken root.

Almanacs & Groundhogs

The chief purveyor of Old World weather wisdom was and is the ever-popular North American weather almanac. Not that almanacs were a New World invention. England had them as early as Shakespeare's time. Keith Thomas in his *Religion and the Decline of Magic* states that William Lilly's *Merlinus Anglicus*, a yearly compendium of everything from quack medicine to weather lore, sold

THE EVER-POPULAR FARMER'S ALMANAC

18,500 copies in 1648. In 1670, people swore by *The Shepherd of Banbury*, a collection of rules and maxims by John Claridge. These almanacs were a welcome diversion in a world dominated by the Bible, the Book of Common Prayer, religious tracts and handbills. (*Pilgrim's Progress* was still eight years away.)

So when people left for the New World, they packed their almanacs as carefully as their Bibles and kept right on using them. Gradually, though, New World weather wisdom infiltrated the old beliefs, and almanacs became a blend of both.

Rube Hornstein, chief meteorologist at the Dominion Weather Office in Halifax for many years, had little use for almanacs except for a good laugh. He never tired of poking fun at their inherent quackery and silliness. In *Forecast Your Own Weather Book*, he compiled a series of English maxims based on saints and proved that, if you took them all literally from Saint Swithin's Day to December 31, rain would fall every day of the year! He also pointed out that, since many of these sayings dated from the Middle Ages, the adoption of the Gregorian calendar in 1582 would have put most of the dates out of kilter anyway.

Hornstein considered almanacs a harmless and amusing shell game played by tongue-in-cheek editors who always left themselves a loophole. Flick through any current weather almanac and note the same hedging tone. "Expect snow on these dates...", they say, confident they'll be right half the time, which is enough for most readers. Hornstein even deflated our most beloved weather myth, that of Groundhog Day, February 2. As the old rhyme goes,

If Groundhog Day be fair and bright,
Winter will have another flight;
But if Groundhog Day brings cloud and rain,
Winter's gone and won't come again.

He pointed out that the animal in the original rhyme was the European hedgehog, a small insect-eater which winters underground but is apt to emerge on any mild day. Here's the original version (which likewise refers to February 2):

If Candlemas Day be fair and bright,
Winter will have another flight;
But if Candlemas Day brings clouds and rain,
Winter is gone and won't come again.

The New World having no hedgehogs, the pioneers substituted the much larger, herbivorous woodchuck—and a legend was born. Hornstein, on his own radio program "Meet Your Weatherman," tried hard to debunk it. But, as much as Maritimers esteemed their weatherman, they loved the rolypoly 'chuck even more. None more so than Nova Scotians, who for years have looked to Shubenacadie Sam, a resident of the provincial wildlife park.

But this isn't simply a Maritimer fixation. Witness the fuss in Ontario a few years back when weather mascot Wyarton Willie expired. With Groundhog Day only days away, the mayor and councillors were biting their nails. What were they to do about Willie's annual pronouncement? There simply wasn't time to find a substitute. There seemed to be only two options: do nothing, or....

Fortunately, someone had thought to freeze Willie's corpse. So Wyarton's worthies thawed the rodent out and got a taxidermist to stuff him. On February 2, the resurrected Willie was photographed emerging from his den. With his two glass eyes, the rodent looked a trifle stiff and dazed, but emergent groundhogs are never that athletic or vivacious anyway. The ruse seemed to work. But then someone—a child no doubt—noticed that the groundhog wasn't breathing. The councillors came clean; the dust slowly settled. Wyarton installed a more youthful Willie, Ontarians felt better, and the well-meaning fib was history.

With a warm winter sun beaming down, Shubenacadie Sam finally made his appearance to forecast another six weeks of winter yesterday.

—*Truro Daily News*, Feb.3, 1999

Weather Signs

If we can't trust old British epigrams, or modern almanacs, or the prognostications of groundhogs, who or what can we trust? Fortunately hundreds of homegrown weather maxims have come down to us. And while many don't hold much more water (pardon the pun) than the foreign maxims, some can be trusted. Mr Hornstein himself trusted a few of them, at least for short-range forecasting. He had seen enough to know that local experience is at least as reliable as science.

Among the maxims he didn't trust happens to be a favourite of mine. It's the one about predicting the severity of the coming winter by the width of a woolly bear caterpillar's stripes. Most Maritimers have seen this little critter, larva of the tiger moth, hustling along in late summer in its black fur coat and rust-red sash. The wider the sash, they say, the milder the winter. Anything so cute, you just *want* to believe. Eric Sloane, America's first TV weatherman and a rural New Englander, reckoned there was something to it.

Narrow band = "Mild"

Wide band = "Cold"

WOOLLY BEAR CATERPILLAR

But Hornstein didn't agree. Gleefully, he cited the autumn of 1952, when the caterpillar's narrow belt supposedly foretold a severe winter. On the strength of this, plus high hornets' nests, heavy berry crops and the like, Maritimers had invested in extra snow shovels and weather stripping, bought ear muffs and mittens, and hunkered down for the worst. The winter was one of the mildest in years.

Still, Hornstein did give a cautious nod to the pimpernel, (*Anagallis arvensis*), a small red, white or purple European flower whose petals are said to fold before rain. Studies have proven they do—if relative humidity exceeds 80 percent. Nova Scotians called it the Poor-Man's-Weather-Glass.

Humidity is the experience of looking for air and finding water.
—Fifth-grade student

Hornstein also appeared to put some faith in foretelling summer weather by the sequence in which certain kinds of deciduous trees leaf out. The rhyme in question goes like this:

If the ash is out before the oak
You may expect a thorough soak;
If the oak is out before the ash,
You'll hardly get a single splash.

But then he impishly quoted another version that flatly contradicts this.

The belief that swallows and bats fly low just before rain got his nod because it makes scientific sense. Both creatures feed on flying insects, and insects do fly lower as atmospheric pressure falls. It may be that buoyancy is higher there. And in summer a falling barometer is usually a sign of rain.

ASH OR OAK?
"HARDLY A SPLASH," or "A THOROUGH SOAK"

Some other sayings which Hornstein and/or Sloane agreed with are:

Chickens take shelter before rain.
Horses and cows act up before a storm.

When cats get restless and wrestle, there'll be a change in weather.

Ants scurry about before rain; spiders are busy repairing their webs, and fireflies emerge in greater numbers.

"Sea gull, sea gull, sit on the sand; It's a sign of rain when you're at hand."

Frogs and toads come out just before rain.

Cobwebs on the grass are a sign of frost.

When human hair becomes limp, rain is near. [This is what makes those Old-Man-and-Old-Woman weather devices work; dry air shortens the hair, making one figure emerge; damp air expands it, reversing their positions.]

When smoke descends, good weather ends. [Low pressure air is less buoyant.]

Cold wet May,
Barn full of hay.

Telephone wires hum and whine when a weather change is due.

When the leaves show their undersides, it will rain.

When conifers glisten and the wind is northeastern, snow will soon come.

And so on.

Like Mr Hornstein, Eric Sloane did his best to debunk certain old sayings that he considered baseless or stupid. His *Folklore of American Weather* labels the following as false:

Dry May, wet June.

When ladybugs swarm
Expect a warm.

Wet June, dry September.

Kill a beetle and it will rain.

Lightning is attracted to mirrors.

Thunder curdles cream, lightning sours milk.

Green Christmas, white Easter.

A rainy wedding day
Makes the skies of marriage gray.

It is bad luck to point at a rainbow.

When squirrels lay in a big store of nuts, look for a hard winter.

Mare's tails [long high wispy clouds] make tall ships carry short sail
[i.e., they prepare for high winds].

That last one puzzled me, for I'd often seen it come true. On double-checking, I found that Sloane hadn't debunked it outright; he merely considered it unreliable.

Maritime Lore

Confusing, isn't it? The trouble is, between the bona fide and the bogus there's a trunkful of heirlooms too charming to part with. Most came out of farming and fishing. The following are mostly Maritime in origin, and all are anonymous.

Onion skins very thin,
Mild winter's coming in.
Onion skins thick and tough,
Coming winter cold and rough.

When the sun is "drawing water," rain is due in a day or so.
[Oldtime sailors called these long downward rays "backstays on the sun," from the stays or ropes that brace a ship's mast.]

When a horse yawns it's a sign of softer [milder] *weather.*

If the breast bone of a cooked turkey is half brown and half white,
then winter will be cold at first and then mild.

When a rooster flies to a high perch and crows in the day, he foretells a storm.

If a dog eats grass it will rain the next day.

Bushy tails on squirrels mean a cold winter ahead.

When muskrats build their shelters early, a tough winter will follow.

Kill a spider, it will rain within twenty-four hours.

The longer a porcupine's quills in the fall, the snowier the winter ahead.

Coyotes yipping and howling mean bad weather on the way.

Ants scatter, expect fair weather;
Ants in file, storm in a while.
Flies and mosquitoes
Biting and humming,
Swallows fly low
Rainstorm coming.

Cows huddle before rain.

A glassy sea with an oily yellow sky soon brews a storm.

A sunshiny shower won't last an hour.

Rain before seven, stop before eleven.

When the wind is in the east
'Tis good for neither man nor beast.

As a country person myself, I've come to trust certain weather signs more than others, at least for my neck of the woods. Here are some I swear by:

If maple sap runs faster, it's going to rain. [Falling air pressure causes the tree's inner sap pressure to be relatively higher.]

Bees stay close to home before rain. [A direct hit by a raindrop can drown a bee, many flowers close before rain, and low-pressure air is less buoyant.]

Fish leap before a storm. [To catch the low-flying insects.]

The more types of clouds, the greater the chance of change.

Catchy drawer and sticky door,
Coming rain will pour and pour.

The sharper the blast
The sooner it's past
[Steep weather fronts usually pass quickly.]

A rainbow afternoon
Good weather coming soon.

Morning frost or dew or fog
No rain today will you log. [Both usually follow a clear night.]

Dew on the grass,
Rain won't come to pass. [Same idea.]

Dew-laden spider webs and grass at dawn mean a fine day ahead.
[Ditto.]

Counting the number of cricket chirps in eight seconds and adding
four will give the temperature within one degree Celsius. [Insects
have little or no internal heat, so their functions speed up or
slow down with changing temperature.]

When ditch and pond offend the nose,
Look out for rain and stormy blows.

When the chairs squeak
It's of rain they speak.

It's too cold to snow. [Since snow is formed
when moisture-laden air is chilled below
freezing, the heaviest snowfalls occur near 0°
Celsius or 32° Fahrenheit.]

April showers bring May flowers.

Worms come up from the earth before heavy
rain.
[If they don't they'll drown.]

"APRIL SHOWERS BRING MAY FLOWERS"

Snow like meal, snow a great deal;
Snow like feathers, softening weather.

When down the chimney falls the soot
Mud will soon be underfoot.

Aches and pain—
Coming rain.

Flowers smell sweeter before rain.

Wind from the west
Weather is best.

Raindrops lingering on leaves and twigs for an hour or two after rain has stopped mean more rain soon.

If a fiddle (or a guitar) won't stay in tune, there's going to be a storm. [Changing air pressure alters tension by expanding or contracting wood.]

Here's a Scottish version of "Red Sky in the Morning":
Evening red and morning grey,
That's the sign of a bonny day;
Evening grey and morning red,
Ewe and lamb go wet to bed.

The rain isn't over till the pavements are dry.

Birds stock up at home feeders before a storm.

When you see the new moon with the old moon in her arms [i.e., when the shadowed and usually invisible side of the new moon can be seen], *bad weather is coming.* [Earthshine from massed clouds on the sunlit side of the earth causes this.]

Stoves draw better during high pressure and poorer during low pressure.

Here are a few more I've found to be mostly true:
When sailors sees the land "looming" [seeming to lift off the horizon] *they head for port because a storm is brewing.*

[Colchester County's Five Islands are noted for this phenomenon; so are Shelburne County's Baccaro Point and many other headlands throughout the region that can be seen from afar.]

When wood fires suddenly spark, or if sparks [often called "soldiers" or "wild geese"] *march around the sooty edge of a stove lid hole, rain or snow will soon follow.* [Oldtimers took it further and said it foretold a funeral procession.]

When rubber tires pick up pebbles along a gravel road on a frosty morning, a thaw is coming.

Rocks coated with ice under flowing water mean soft [mild] *weather on the way.*

Northern lights flickering against the winter sky mean a thaw or rain.

A sky glittering with many stars foretells bad weather; few stars mean a fine day. [Stable air grows hazy with moisture and dust; disturbances are usually preceded by high-altitude cold air that sweeps the haze away. Likewise, fuzzy horns on a crescent moon mean stable air; sharp horns portend change.]

Sun hounds or sun dogs [rainbow-like spots on one or both sides of the sun at evening] *foretell bad weather soon. If only one side shows a spot, the storm will supposedly come from that quarter. A ring around the sun or moon means the same thing; the larger the ring* [i.e., the lower the cloud deck], *the sooner the change.*

When the wind backs against the sun, trust it not, for back 'twill run. [For example, if the wind shifts from east to northeast to north (counter-clockwise), it will do the same in reverse, often bringing the rain or snow back. But if it follows the normal clockwise sequence from, say, east to southeast to south, it will likely continue until it becomes a westerly, bringing fine weather.]

One of my favourites comes from Tom Parker's delightful *Rules of Thumb*, a collection of commonsensical sayings from all over North America. School teacher Lin Spaeth observed,
When first-graders get disruptive in class, there's going to be a major change in weather.

With so many weather sayings to choose from, it's a pity we have so few from the northeastern native people. Surely in thousands of years of living here they would have coined the most dependable maxims of all. In my limited research I found these few, unfortunately not from the Maritimes:

When the buffalo band together, the storm god is herding them.

When the sun sets unhappy [i.e., with a veiled face], *the morning will be angry with storm.*

When you can hang your powder horn on the moon, do just that. [It means dry weather and therefore noisy stalking and no game for the pot.]

When the night has a fever, it cries in the morning. [A rise in temperature before midnight leads to rain.]

16

Cloudscapes

Remember how as children we used to lie on our backs and watch flotillas of clouds sail overhead in ever-changing shapes? Science, not blessed with childlike humility, has long been able to explain clouds. Yet they still speak to the child in us. Thanks to the movie camera, we've discovered that their slow-motion shapeshifting, almost too gradual for the unaided eye to see, is full of startling energy and grace. Puffy shapes appear, fling out amoeba arms, twirl like ballerinas, and dissolve.

"It is like a flower being born," wrote American weathercaster and cloud painter Eric Sloane, "pushing up white petals the size of apartment houses, writhing in almost animal ecstasy till it has reached the full stature. Its flow back into the air is an equally fine spectacle."

Sloane's lyrical description moved me to see a cloudbirth for myself. Selecting a bright and breezy September afternoon, I flopped on the grass and stared into the blue vault. The first thing I noticed was that the sky overhead is surprisingly dark—almost violet in fact. That's because I was looking through the thinnest part of the atmosphere. Here the scattering of blue light, which makes the sky look progressively bluer and paler toward the horizon, was minimal. The blackness of outer space was showing through.

I am not sure how clouds get formed. But the clouds know how, and that's the important thing.

—Fourth-grader

Directly overhead, startlingly white against the dark void, several white cumulus clouds could be seen. They seemed to be stationary, like ships at their moorings. I decided to spy on one for a while. To my surprise it seemed to unravel, then passed out of sight over the eave of our house.

Next I noticed, near the horizon, a triangular scrap of cloud. It seemed to be twirling and growing, but so slowly I couldn't be sure. After waiting a moment with eyes shut, I found that the cloud had tripled in size and rolled over. This method of time-lapse looking enabled me to see cloud mechanics—if so hard a word can be used for so soft a thing—that I'd never noticed before. Minutes later my cloud galleon began shrinking like a snowflake in the spring sun, and soon vanished. Not long after, the same bit of sky gave birth to three or four versions of the vanished cloud, all smaller than the first.

Clouds & Emotions

Of course Mitchell wasn't singing about real clouds. Moody like us, clouds make ideal symbols for mental states and for philosophical insights.

Clouds also affect our emotional states. From earliest times those floating mist cathedrals have tickled our imaginations and inflamed our fears. A week of overcast sky gets us down. An evening sky ribbed with horizontal clouds is calming. A full moon looks more mysterious partly veiled in cloud. A blue sky with high cottony clouds cheers us up. Children of all ages delight in seeing fantastic animals in clouds, some comical ("That one looks like a clown!"), some menacing ("See the shark?"). One child defined clouds as "high-flying fog."

Clouds cause physical discomfort as well. While they create glorious sunsets, they also harbour lightning that can blast trees and people with forked fire. While they adorn the plainest landscape, they *do* rain and snow on everything.

Unlike our songwriter, meteorologists make it their business to know clouds inside and out. They tell us that clouds come in three basic types: *cumulus* (lumpy), *stratus* (layered) and *cirrus* (wispy).

Watch a smoker puffing on a cigarette, and you'll see the first two. Each separate new puff is like a cumulus cloud. If he smokes long enough and the room isn't drafty, a stratus cloud will form along the ceiling. And you can see cirrus-like clouds when you open the deep freeze door and wisps of chilled vapour curl out. Add the prefix *alto* (middle) and you can classify fur-

"The sky," said the New England philosopher Ralph Waldo Emerson, "is the daily bread of the eye." It's mostly clouds that make it so. Without them every sunrise and sunset would look much the same.

I've looked at clouds from both sides now,
From up and down, and still somehow
It's cloud illusions I recall
I really don't know clouds at all.

—Joni Mitchell, "Clouds"

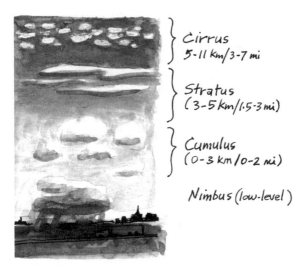

Cirrus
5-11 km/3-7 mi

Stratus
(3-5 km/1.5-3 mi)

Cumulus
(0-3 km/0-2 mi)

Nimbus (low-level)

**THREE BASIC CLOUD TYPES
(PLUS ONE)**

ther, as in *altostratus*. Add *nimbus* (head) to cumulus or stratus and you have two more types. Combine these, as in *nimbostratus* and *cirrocumulus*, and you have still more.

Different cloud types suggests different weather outcomes. Thus, cumulus is a fair weather type, nimbus comes with rain or snow, and stratus is an intermediate type.

How do clouds form in the first place? Why aren't there always clouds in the sky—since even over a desert there is moisture in the air? The explanation is that warm, dry air can hold much more water vapour than cool air without showing it. The microscopic vapour droplets are too few among the molecules of nitrogen, oxygen, carbon dioxide and trace gases to be visible. We might compare this to moisture in a damp towel. It's there, we can feel it; but we can't see it. Wring the towel, however, and the water droplets appear.

Cool or cold air is denser, takes up less space. But water cannot be compressed. As lower temperatures "squeeze" rising air, vapour particles condense; i.e., they clump together and form larger droplets. If they keep rising and cooling and condensing, they finally grow too heavy to float, and fall as drizzle, rain, or in freezing temperatures as ice pellets or snow.

The temperature at which the moisture in a given parcel of air condenses and becomes visible is called the "dew point." When you open the bathroom door onto a cool hallway after a hot bath, steam momentarily appears. Inside the bathroom it remains invisible because the warm air absorbs it all. But the cooler hallway air can not, so the vapour condenses. It doesn't

The amount of water in a cloud is minuscule. A cloud the size of a large iceberg might contain the equivalent of a full bathtub. Less than five percent of the world's water is tied up in clouds.

DEW POINT

fall to the floor (rain) because the air isn't *that* cold and the amount is too small. But it may coat the nearby walls and door.

We get a similar effect as we breathe out on a cold winter day. The warm air from inside our wet lungs is suddenly compressed, revealing its cargo of water vapour. Should the outer air be nearly saturated with moisture, which provides nuclei for condensation, our puff of steam will persist longer.

Most clouds become visible only at higher altitudes, but one type forms at ground or sea level; we call it fog. Fog is identical to the clouds whizzing by the windows of an airborne jet, only not as cold.

Night drivers often find themselves suddenly wrapped in white mist, usually on a cool, windless night under clear skies after a warm, damp day. As the earth radiates its daytime warmth into space, the chilled moist air, now denser, flows downhill like water. As it pools in hollows and valleys it becomes visible as white "radiation fog." This is common in mountain valleys at dawn. The last time I flew from Vancouver to Halifax, every valley was full of mist; the Rockies seemed to float above a white plain.

Another kind of fog occurs when warm air slides over cold water (sometimes called "sea smoke" in winter) or over chilled level ground. Our most common fog is caused by warm, wet Gulf Stream air sliding over water chilled by the Labrador Current. I've seen the same thing on a smaller scale while driving across the Isthmus of Chignecto, where warmish Northumberland Strait air mingles with cool Cumberland Basin air.

RADIATION FOG

Such fog often forms along the Eastern Seaboard, especially on warm summer nights. On fine days our prevailing westerlies, helped by the sun's heat, usually nudge it offshore by mid-morning. Sailors see a lot more fog than landlubbers do. The foggiest region is over the Grand Banks northeast of Newfoundland, where the two great ocean streams collide and mingle at full volume.

But let's return to the sky. What generates clouds in the first place? Oddly enough, it's the presence of a heat source on the land or sea below. This warm spot can be a patch of shallow, sun-warmed water, a stretch of dark plowed earth, a newly paved parking lot or an airport—anything that absorbs and holds the sun's heat. Conversely, snowfields and white sand beaches make poor cloud machines.

Warm spots heat the air above them, lofting the warmed, lighter air, which rises like smoke from a chimney. These balloons of air, on reaching the dew point for that day, become visible as white mist—fine water droplets suspended in air. As the heating continues, the area lofts more and more warm air, and the resulting cloud expands like a giant cauliflower. Eventually it casts a shadow that shuts down the heat source and may reverse the flow. But if other sources between the cloud shadows keep on generating updrafts, alternating patterns of updrafts and downdrafts develop.

By midmorning or early afternoon the sky may be polka-dotted with flat-bottomed, puffy, white cumulus clouds. The higher they rise, the drier the air and the less chance of rain. How high they rise depends on the temperature up there, which partly depends on the season. Winter clouds tend to form at lower elevations than do summer clouds because the dew point—marked by the flat bottom—is lower.

Meteorologist and artist Eric Sloane recounts how, as a novice painter of cloudscapes, he once put summer clouds in a winter landscape. A weather-wise friend promptly corrected him, and he never forgot the lesson.

Of all the cloud types, none is more impressive than the thunderhead or cumulonimbus. It usually forms in the afternoon during hot weather when a cold front—a sloping wall of cool air—plows under warm air, setting up violent vertical air currents that tear water droplets and their electrical charges apart. This creates different charges in the upper and lower zones of the cumulus tower and between it and the earth.

Eventually such a cloud tower will rise above the freezing zone (about 4km/2.5mi in summer), and develop the classic flat-

The infamous pea-soup fogs of London, England arise when moist Atlantic westerlies meet the city's notoriously polluted air. The particles of soot and other impurities attract the water droplets. Reduced pollution in recent decades has reduced the duration and severity of pea-soupers.

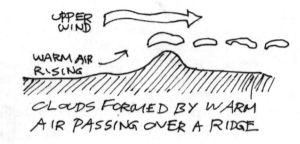

CLOUDS FORMED BY WARM
AIR PASSING OVER A RIDGE

tened (anvil) head, often leaning downwind. If the negative
charge at its base becomes great enough, electrical discharges
will equalize the imbalance with a lot of noise—a thunderstorm.
(See Chapter 18.)

Altostratus clouds form in dry air that has a low dew point.
This is typically above 3km/1.8mi altitude, just under the zone
where all moisture is in the form of ice crystals. Often they form
the leading edge of a warm front creeping in from another
region. As the lighter air of the front rides up over the denser,
cooler air, the ceiling will lower and darken until it becomes a
thick, ragged nimbostratus cloud threatening rain within twenty-four hours. Such a rain may last for several days. As the old
rhyme has it,

Rain long foretold, long last,
Short notice, soon past.

Cirrus clouds never descend below the freezing zone. They
sometimes form wavy patterns which meteorologists call cirrocumulus. Their resemblance to fish scales gives us "mackerel
sky." Higher than altostratus, they often foretell windy and perhaps warmer weather.

Can any man read the secret of the sailing clouds
Spread like a carpet under his pavilion?
See how he unrolls the mist across the waters,
and its streamers cover the sea.
—Book of Job 36:29-30 NEB.

Clouds are shape-shifters. They don't look the same very long.
The various types can morph into one another and back in the
space of a few days. Interestingly, most clouds are homegrown,
not imported. Cumulus clouds generally disappear overnight
(when the heat source cools) and reappear next morning if it's

sunny. One can think of clouds as moisture made visible or—since the atmosphere always contains *some* moisture—as moisture made more visible.

Dew is formed not by fog near the ground, but by warm moist air depositing its water on chilled surfaces. If the temperature is low enough, the dew will turn to frost.

Put it all together—the many cloud types; their constant motion recalling great sailing ships; their vast size; their habit of celebrating with rain, snow, hail, and sleet, but also with spectacular dawns and sunsets, not to mention lightning and a dozen other light effects—and we surely have one of nature's most wondrous phenomena.

17

Fire from

the Sky

*The storm had rolled away
to faintness like
a wagon crossing a bridge.*
—Eudora Welty

The summer we were struck by lightning our youngest child was about seven, so it must have been around 1983. The bolt set fire to our basement, and afterward the fire chief told us that the house wouldn't have lasted half an hour if we'd been away at the time. Half an hour. You don't soon forget a thing like that.

Like most nineteenth-century Maritime farmhouses, our place was and is eminently combustible. Framed with rough hemlock, floored with resinous pine, finished inside with plaster over spruce laths, clad outside with spruce clapboard, it was lucky to get past the century mark unscathed. The 1940s addition of asphalt roof shingles only increased the risk.

An attic fire would have been bad enough; the basement fire could have torched its way up through all that tinder in minutes. For weeks afterward I'd snap awake at 2:00 or 3:00 A.M., heart pounding from another flaming nightmare. At last I understood why Coleridge's Ancient Mariner kept telling and retelling his tale of woe to anyone who would listen:

Since then, at an uncertain hour,
That agony returns:
And till my ghastly tale is told,
This heart within me burns.

I was like that for months. Telling seemed to calm my fears. The lightning strike itself was nothing much; just a blinding flash and an earsplitting *Crack!* directly over our roof, with no pause between. We ourselves weren't hit, so I figured the house was okay too.

That particular thunderstorm had been building all afternoon. And because such storms are common here at the head of the Bay of Fundy, we didn't pay much heed. The converging shores of Cobequid Bay seem to funnel them through here. One can observe the process any sunny afternoon from July through October. As the sun warms the south-facing slopes of the 300m/1,000ft Cobequid Hills to the north, it lofts balloons of moist, warm air. Cooling, they become visible as towering white cumulus clouds. When their tops bump against freezing air five or six kilometres up, they flatten into the familiar anvil shape. Should an influx of cool air herd several of these columns together, a thunderstorm may happen.

We noticed nothing unusual until around 4:00 P.M., when there was a distant mutter of thunder and the windows darkened. First, the muting of birdsong. Then a brisk wind tossing the birches and hitching the weathervane from west to southwest. Strobe-like flashes growing brighter. The first raindrops rattling on the steel eaves-troughs, to be drowned out moments later by the escalating racket. Just a summer thunderstorm doing its thing.

Half an hour later the rain slackened, the thunder moved on, the trees straightened up and fixed their hair. Sequins of sunlight glistened from every leaf and blade of grass like brushstrokes in a Constable landscape. Songbirds resumed their conversations. Overtaxed downspouts gurgled, and spat out gobs of what looked like coffee grounds but were really bud scales and twigs from spring leafout. The air was perfumed with mud and ozone.

Like I say, we weren't alarmed. I did fetch some wettables and breakables inside. My daughter and I were the only ones at home, and we just watched and listened, walking hand-in-hand from room to room like kids at a fall fair, enjoying the fireworks while keeping a safe distance from the windows. At one point she said proudly, "Daddy, I'm not scared of lightning any more!" Still, when that awful blast shook our roof, her hand momentarily tightened on mine—or was it mine on hers? Five minutes later, we were again enjoying the show.

Until, that is, the awful moment when we smelled wood burning. We were near the kitchen window, watching the retreating flashes twinkle against the far blue hills, when she turned to me and said, "Daddy, I smell smoke."

So did I. But how could it be? In fourteen years of living here, we had weathered many such storms, some much worse. Even though our farmhouse sits on a broad, drumlin-like hill, one of the highest around, with two tall trees nearby, to my knowledge it had never been struck by lightning. Nor had our neighbours' houses. And we hadn't lit a fire in the woodstove since May, or used the oil furnace in weeks.

A quick round of the rooms revealed nothing amiss. The odour seemed to have vanished—or maybe our noses were getting used to it? Telling Joyce to stay put while I checked the basement, I flicked the cellar light switch. Oh-oh. The power was off. Grabbing a flashlight, I was halfway down the cellar steps when I smelled smoke again. Seeing no flames, I took it for ozone from the blast. And, not wanting to leave my little girl alone upstairs too long, I gave the cellar a cursory glance and jogged back up. "There's no fire," I told her.

However, I did check the phone. It too was dead. Strange, for it had never failed during other storms.

Fifteen minutes later the smell of smoke was undeniable. Nervously we checked the ground floor. Nothing unusual in the kitchen. Everything normal in the family room, where our old woodstove stood cold and innocent in its coat of summer polish. And nothing amiss in the central hallway, its high windows bottle-green under the great elm's canopy.

But the living room! The beautiful pine floor was ankle-deep in white smoke that unfurled in slow motion from the furnace registers like the poison gas it was.

My mind raced. Somewhere below us a fire was burning after all. Big or small, I couldn't tell. Probably small, or I'd have seen it, even with the flashlight on. By now it could be well along, though. I listened for the telltale crackle of burning softwood: nothing.

Firemen always warn you to flee a building the moment fire is detected. But how could I abandon our home without being sure? My spouse was ten kilometres away with the car, helping at a church event. She might not be back for an hour. For me to run next door with a little girl in tow would waste precious time. Especially if their phone was also out, as it might well be if a transformer had been hit.

No, first I had to know what we were up against. And if there *was* a fire, the least I could do was try to slow it down.

I told Joyce we'd be okay (about the house I wasn't so sure), and stationed her near the porch door where we could grab stuff and run. I took the big ABC fire extinguisher from its bracket and made for the basement again. It's funny how a person's mind works at such a time. On the surface my thoughts veered between dread and excitement, yet at a deeper level I was coolly listing what to rescue: personal papers, albums, silverware, yes; paintings, books, yes. Clothes? Maybe....

Back down in the basement, the acrid odour of burning wood was unmistakeable. This time I headed directly to the windowless low back part where the furnace sits above an earthen floor. Overhead, the flashlight's beam lit up more white smoke. It covered the joists like creeping mould. It was coiling around the heating pipes and dropping tendrils that weaved to and fro like blind worms.

Still seeing no actual flames, I clicked off the light. As my eyes adjusted to the dark, a flickering orange glow appeared behind the furnace. Keeping my head below the fumes, I approached. The fire was about a metre wide, silently licking along two joists that nearly touched. It was barely burning because the trapped smoke was starving it of oxygen. I ripped the seal from the extinguisher, took aim, and squeezed. Instantly the glow winked out and darkness swooped back.

Throat and eyes burning, I retreated upstairs. Anyway, I consoled myself, the fire is out. Now to get rid of the smoke. We were congratulating ourselves and getting ready to open windows and doors when I noticed an ominous new spot in the middle of the living room's varnished floor. About the size of a hen's egg, it was coal black in the middle, shading delicately to toast-brown around the edges. And it was too hot to touch.

So the enemy below was still alive. I recollected that when we renovated the house, the carpenters had left some old joists to spike the new ones onto. Those old three-by-eights, exposed to dampness in the years before we bought the place, must have developed dry rot. If so, they would be punky and fibrous by now, the perfect nest for a smouldering fire. Not only that, the fire must have broached the sub-floor and be eating into the pine planking above, or there'd be no scorch mark.

So my burst of ABC powder had only killed the surface flames. ABC extinguishers aren't designed to cool things anyway. Water is the best for that. Without water the wood can smoulder like tobacco in a smoker's pipe for hours, bursting out the moment it tastes oxygen.

So the stupidest thing I could do would be to open doors and windows. Stupider still was my wild notion of cutting out that black spot with my chain saw.

Sweating now, I grabbed our small ABC extinguisher, pounded down the steps, and again sprayed the joists until the pressure faded, which didn't take long. Water. My kingdom for a pail of water! But where to get some, with the power off and our well pump dead?

Of course! The toilet! It must hold 10 or 15 litres at least. All I needed was some sort of open container, a mug maybe? No, too small...an ice cream tub? Yes, good, I could slosh water up there, and....

My thoughts returned to that widening scorch on the pine floor. What if the fire hit that varnish before I could douse it from below? My equipment was poor, the visibility almost nil. Better to use my hoard of water to soak a mat and lay it over the hot spot, and to keep on wetting it until the water gave out or Beth got home, whichever happened first.

Luckily, Beth's arrival came first. As soon as I told her what had happened, she drove off to phone the local volunteer fire brigade. Within ten minutes their scarlet and chrome truck arrived in our yard, siren wailing. Minutes later the fire was out, and a big fan was vacuuming the smoke and steam away.

They determined that the lightning bolt had entered the house through the telephone box on our northeast wall. From there it had travelled along the thin copper wire until, three metres in, the wire entered a hole bored through the two old joists. At that point the copper strand had shorted out, igniting the dry wood.

Around 11:00 P.M. the chief, having checked the nearest transformer and turned the power back on, sent most of his crew home. Not all, for if the lightning had damaged our wiring too, a second fire could start inside the woodwork hours after they left. Not until 1:30 A.M. was he satisfied to go.

Satisfied or no, that night I slept on the living room floor next to that black mark. And every half-hour I fingered it to make sure it wasn't hot....

The reality of all this didn't sink in until next morning. A bolt from the sky had very nearly destroyed our home and everything in it. Moreover, had Joyce or I been standing in the wrong place, say under the tree or next to the telephone box, we could have been killed.

The telephone official doubted that lightning could follow a phone line indoors like that. It was theorized that a forgotten buried wire may have provided a channel from tree to box. (I've since confirmed that lightning does sometimes follow telephone lines into buildings unaided.)

Anyway, it happened. Checking the basement ceiling afterward, I was able to poke my three fingers through the hole in the sub-floor and halfway through the top board. Another centimetre and the fire would have broken free.

The insurance company offered us six hundred dollars to repair the floor. The old pumpkin-coloured pine board was irreplaceable, so we opted to keep the scar as a memento of our narrow escape. It's still there, with the house still attached.

18

When our firstborn son was a teenager working as a farm hand, he liked to play a trick on urban visitors. In the course of showing friends around, he would point out the single strand of electrified wire which kept the cows from running away. Right on cue, the visitors asked whether it hurt to touch the wire. "Naw," he'd say. "It jolts the cows some, but if they wore boots or shoes like you and me it wouldn't bother them at all." Which was more or less true. Then he'd walk over, grab the live wire, and stand there grinning.

"What's it feel like?" they'd ask.

"Oh," he'd say, "just a nice little buzz. Here, see for yourself. The voltage is very low."

Well, if this hick in a green plaid shirt could do it, so could they! Gingerly one of them would touch a finger to the wire. *Whap!*

If they wouldn't try, he'd offer his other hand, telling them to press one hand into the grass to "lessen the shock." Which was a bald-faced lie. As every farm kid knows, electricity always heads for the ground. The secret, he explained, was to grasp the wire firmly with one's whole hand. That way the current wasn't forced to travel though one finger, causing the jolt. But Danny would get no more takers that day.

Lightning has been rightly called one of nature's gravest hazards. It has killed or maimed people as they fished from metal boats or cycled down a street or played golf or vacuumed the rug or fiddled with a TV dial. It has killed cows and horses as they sheltered under trees, racing along the roots and up their wet legs.

Thor's Hammer: A Peek Inside Thunder & Lightning

In midafternoon...
a curious darkening of the sky,
and a lull in everything that had
made life tick and then
the way the boats suddenly swung
the other way at their moorings
with the coming of a breeze out
of the new quarter,
and the premonitory rumble.
Then the kettle drum, then the
snare, then the bass drum and
cymbals, then crackling light
against the dark, and the gods
grinning and licking their chops
in the hills...
—E.B. White, "Once More to the Lake" in *One Man's Meat*

Awesome Power

The power of a lightning bolt is stupendous. A single millisecond flash unleashes millions of volts. Electric chairs require only a few thousand volts to send their victims to the great beyond. Hundreds of amateur electricians have been killed by standard 120-volt house wiring. Even a 12-volt automobile battery packs a nasty wallop.

Our mischievous son wasn't being sadistic, or playing at shock therapy. He just liked the physics of it all. Lightning fascinated him, too. The physics of lightning is that it always takes the easiest route to ground, the path of least resistance. Almost any conductor will do—tree, post, hydro pole, cow, human. If the conductor is wet, so much the better. If it's metallic, preferably something dense like copper, silver or gold, better still.

In the sky, where there are no such conductors, lightning tracks the spiralling interfaces between ascending and descending air, the movements of raindrops and hail. At the moment of passage its zigzag path heats the air to as much as 30,000°C/54,000°F, five times the sun's surface temperature, enough to vaporize the hardest metal. By contrast, all it takes to ignite dry wood is about 250°C/480°F.

When our maple was struck, it was lucky to suffer no more than a split trunk. Actually the trunk had been split years before, whether by lightning or wind I don't know. The bolt widened the old seam while opening a new crack at right angles to it. The tree just kept on growing, but the cracks worried me. My fear was that some night, in a blizzard or a northeaster, its sprawling branches would tear down our telephone and hydro lines and maybe wreck the roof as well. The tree might have to be felled.

Danny, now living in Manitoba, recently described another tree struck by lightning at a church camp he was attending. "A fully healthy cottonwood, 14 inches at butt, [was] hit in the campground—split top to bottom several ways and mostly debarked," he wrote. "A 10-foot-long 'splinter' stuck in the ground 25 feet away—quite a bang." The next evening it rained, and "...the flood brought the river *through* the auditorium, about 15 minutes into the evening service. It was just deep enough to cover the calves of the legs of the people kneeling at the front (one would have to see it to believe it). Both events were thoroughly enjoyed by all."

Theoretically the cottonwood's violent demise could have been prevented by "grounding" it—running a low-resistance wire from top to bottom and hooking it to an iron or steel rod stuck in the earth at its base. While few take this much trouble over a tree, we do so routinely for skyscrapers, power poles, church steeples and other tall structures. Because power pylons are perfect targets, they're fitted with lightning arrestors that automatically switch off the current when a bolt hits, and switch it back on afterward.

Without such protection, structures like Toronto's CN Tower, though made of concrete, would suffer lightning damage all the time. The tower's steel reinforcing rods are all carefully grounded.

Lightning strikes the CN tower roughly two hundred times a year.

Lightning Rods

On the highest points of older structures one sometimes sees vertical rods at intervals, sometimes with ceramic balls. Lightning rods were first proposed in 1749 by the American amateur scientist Benjamin Franklin. His idea was to "catch" aerial electrical charges and convey them safely to ground without harm to the structure itself.

The idea is sound, but to work properly the rods must be tall enough to keep the bolt from also hitting the structure. In theory each rod protects a conical area whose base diameter is twice its height above ground. So a rod twenty metres above ground should shield everything inside an imaginary cone forty metres wide at ground level. Anything outside that cone needs additional rods. Still, lightning has been known to strike inside the so-called safe cone.

One of the ironies of my own lightning episode was that only a year or two before, I had refused a lightning rod salesman. He had come while I was working in the garden. "How much?" I asked, sticking my fork in the dirt.

"Seven-fifty a rod, installed," he said.

"Seven dollars and fifty cents?" I asked, never having priced one before.

"Seven *hundred* and fifty," he said.

I'd waved him on with a laugh, secretly hoping my maple would catch it first. It did; but not enough. A lightning rod or two might have saved me some white hairs.

During my university years in Fredericton, New Brunswick, I often walked by a downtown church, Wilmot United I think it was, which had a tall wooden spire. Tall spires on old churches aren't that unusual, but this one boasted a carved wooden hand whose forefinger pointed heavenward—with a lightning rod at its tip.

In 1752 Benjamin Franklin proved that lightning is electricity. He did so by flying a metal-tipped kite from a silk thread attached to a key during a thunderstorm. By touching the key during a flash, he induced a spark and simultaneously felt a shock. Franklin was lucky; the next two people who tried this were electrocuted.

Thou shalt not tempt the Lord thy God!

When Franklin's idea hit Europe there was a storm of controversy about the use of lightning rods. Scientists argued that sticking metal rods on top of buildings would only invite a lighting strike. Indeed, it was widely believed that churches, tall steeples and all, were safe from God's lightning.

In 1767 authorities in Venice decided it would be sacrilegious to suggest that God would allow lightning to strike a church, so they stored hundreds of tons of gunpowder in a church vault. When lightning hit, the explosion killed three thousand people and destroyed a large part of the town. By 1772 such landmarks as St. Peter's in Rome and Venice's Cathedral of San Marco had lightning rods. On the thundery afternoon of April 18, 1777 crowds filled the Piazza of Siena to see if the great tower, recently fitted with the controversial device, would be struck. Lightning did strike—and Franklin was vindicated.

Struck by Lightning!

Every year roughly sixty Canadians and a thousand Americans get hit by lightning. Incredibly, about four-fifths survive. That anyone can survive a force that explodes trees and blasts bricks off buildings is hard to believe.

How does it feel to be hit by lightning? "I don't remember being knocked down," said one victim, a middle-aged computer programmer who had been hit while golfing in 1986. "But I remember I could hear it humming through me. Like that sound you hear at a hydro substation? I tried to move. Couldn't. Finally, I managed to get up...." Though the metal club in his hand made a perfect conductor, he survived.

Another victim, a 28-year-old woman who was struck while working in a garage in August, 1997, knew nothing until she saw a blue flash and heard a loud bang. The bolt burned a hole in one rubber glove and melted the rubber sole of her left shoe. Pain suffused the left side of her body. Her muscles twitched; her legs wouldn't function. She felt stunned and confused, unable to breathe, her mind in a whirl. Her boss rushed her to a hospital, but three years later she was still suffering after-effects. "I see an apple in a store," she said, "and I can't think of what it's called."

Most lightning victims find solace in the company of other victims. She spends a good deal of time on the Internet, helping other lightning survivors, many of whom were less fortunate. "It's a life-changing experience," she says.

Several websites allow lightning survivors to share their stories, among them Lightning Strike and Electric Shock Survivors

Our word "electric" comes from *elektron*, ancient Greek for amber—fossilized tree resin. Around 600 B.C. the Greeks observed that this naturally occurring yellow resin, if rubbed against fur, mysteriously attracted a small feather.

The deep and turbulent atmosphere of the giant cloud-planet Jupiter reveals continuous lightning activity.

International (www.mindspring.com/lightningstrike). Sharing eases the trauma of a close encounter with one of nature's most terrifying phenomena.

Worldwide, an estimated five million bolts strike the earth each day. With two thousand storms in progress at any one time, each producing thousands of flashes, it's a wonder more people aren't struck. Fortunately, detection systems are getting better at predicting when and where the hits will occur.

Forest Fires

Few professionals worry more about lightning than do forest managers. In Canada, thunderstorms ignite four thousand forest fires on average each year, scorching an area half the size of Nova Scotia. Most of these fires occur in the drier central and western provinces, where they far outnumber people-caused fires. (The reverse is true in the moister east.)

Clearly, plotting each strike would be a boon to fire control agencies. Not that all such fires must be put out. Resource managers now accept wildfires as an integral part of many forest ecosystems, especially in the north. Foresters have long known that species like jack pine and black spruce, with their heat-resistant cones, virtually depend on fire for successful germination. Foresters also know that moderate ground fires hasten the composting of bug-killed trees, sanitize the ground after pest and disease epidemics, and clear out stagnant old growth so that vigorous new biological communities can thrive.

But uncontrolled wildfires also consume valuable wood, destroy watersheds, kill trout and baby deer, and ruin hiking, angling, and hunting. And of course they destroy cottages, homes, and even towns.

Given that the earth's most flammable trees are conifers, and that Canada has the second largest coniferous forest in the world, we have a lightning problem. Monitoring lightning fires used to be done by radio-linked lookouts perched in tall metal towers on high hills. However, they could only monitor the more accessible southern forests. After World War Two, bush planes extended the range. Since 1998, thanks mainly to the federal government's $9.5 million Canadian Lightning Detection Network, scientists have been able to pinpoint virtually every lightning strike across the country.

Currently (no pun intended) managed from Halifax, the Network relies on 81 unstaffed stations equipped with sensing antennas, Global Positioning System (GPS) receivers and satellite dishes. Nova Scotia has three stations (Sydney, Yarmouth,

Sable Island), Prince Edward Island and New Brunswick each have one (Charlottetown, Fredericton), and Newfoundland-Labrador has eight (St. John's, Gander, Stephenville, St. Anthony, Cartwright, Goose Bay, Churchill Falls, Wabush Lake).

Whenever lightning flashes anywhere in Canada, four to ten stations detect it and instantly relay data to Global Atmospherics Inc. in Tucson, Arizona. Within seconds, Global's computers locate the strike by plotting where electronic "lines of sight" intersect, and transmit the data to Environment Canada offices.

Similar networks protect the forests of forty countries worldwide. The database they are building is a priceless tool not only

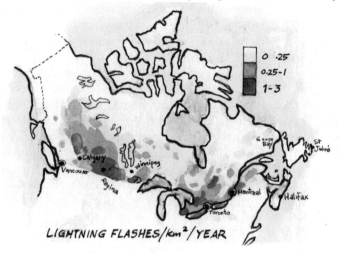

LIGHTNING FLASHES/km²/YEAR

for fire forecasting, but for studying national and global weather patterns.

A case in point is Canada's map of lightning activity. Most of the map—across the top and down both sides—is coloured pale yellow, denoting fewer than 0.25 discharges per square kilometre per annum. Southward the map grows pinker, until in southern Saskatchewan and along the Montreal-Toronto corridor it is tomato-red, representing two to three strikes per square kilometre each year.

Thunderstorm Anatomy

Scientists agree on the basic physics of thunderstorms—convection currents moving warm air up and cold air down, the friction this creates, the buildup of electrons. They also agree that for lightning to occur, there must be a charge-separation great enough to break down air molecules. But they don't agree on what exactly causes such powerful charges. Some say it's the jostling of air currents, some invoke rain and hail, some cite cosmic rays from above.

Whatever the cause, thunderheads hatch copious amounts of positive and negative electricity. And when the differential between oppositely charged zones becomes great enough, a luminous multibranched stream of electrons leaps the gap until harmony is restored. We call this stream lightning.

It's a bit like the charge that jumps between our hand and a doorknob when we scuff across a shag carpet, or the sizzle of green fire in a cat's fur when one rubs it the wrong way in the dark.

Lightning can travel between the top and bottom of a tall cloud, between the base of a cloud and the ground, from cloud to cloud ("heat lightning"), and even from cloud to clear sky. A third of the bolts—the most dangerous third—passes between clouds and earth.

Most of us associate lightning with towering cumulonimbus clouds. Yet it can occur in snowstorms, in sandstorms, above erupting volcanoes and massive forest fires, and even in clear air. "Ball lightning" is of the latter type, manifesting itself as slow-moving beach-ball-sized luminous spheres that flash and flicker out.

Although Ben Franklin clarified the physics of cloud-to-earth lightning, modern scientists have since added insights of their own. For example, that the action is not all top-down. It starts from above, but is met on the way by a positive pulse from below. Specifically, as the invisible "stepped leader" of negative current, descending in discrete, microsecond steps of about 50m/165ft each, comes within 100m/330ft of the ground, a second leader rears up to meet it. At the instant of contact, the ionized channel may be 10m/33ft across. After a nanosecond of darkness, ten to twenty more strokes may flicker along the channel—now collapsed to the size of one's little finger—until the negative charge in the cloud's belly is spent. This flickering "return stroke" lights up the whole zigzag path.

Thunder is the shock wave generated by the sudden explosive heating of the air along the channel. Up close, the sound is an ear-splitting *Crack!* Dampened by distance, it's a low muttering rumble not unlike barrels being rolled along a wharf. Washington Irving, in his tall tale "Rip Van Winkle," likened it to giants bowling in the sky.

Compared to lightning, which travels at 300,000km/sec (180,000 mi/sec), the pace of thunder is glacially slow, a leisure-

Big forest fires like the ones that ravaged Montana in August 2000 generate their own weather, including thunderstorms.

Thunder is good, thunder is impressive; but it is the lightning that does the work.
—Mark Twain, 1908

ly 335m/sec (1,100 ft/sec), the speed of sound in air. By counting off the seconds from flash to boom (try "one bushel of thunder, two bushels of thunder"), one can gauge the storm's distance fairly well.

Safety
Which brings up the subject of safety. With lightning strikes so
rare, why bother with preventive measures? After all, the odds
of being hit are only about one in six hundred thousand....

True, on average, lightning kills far fewer people in Canada
than hypothermia. Still, why take a chance? One strike is
enough. Besides, lightning is five times more likely to injure
than to kill. Who needs impaired vision, sleep disorders, loss of
hearing and memory, tinnitus (ringing in the ears), confusion
and anxiety? Surprisingly, serious burns are unusual.

On July 15, 2000 35-year-old Leslie Roth of Newmarket,
Ontario was struck, not once but twice, while camping on an
island in Georgian Bay. She suffered nothing worse than a bad
headache, but the situation could have been far worse had her
group of kayakers not known what to do. The instant the storm
broke, all eight of them dove for cover in nearby woods, separated, and found a spot away from tall trees. There they sat
cross-legged, each on their life jacket, their feet off the wet
ground. Had they stayed in the open—or, worse, on the water—
or had they clumped together under a tall tree, the outcome
might have been very different.

So the old adage that lightning never strikes twice is false. The
truth is, lightning tends to strike wherever conditions are right.
The tallest object in any open expanse is its single most attractive target.

Early settlers are said to have
planted trees near—but not
too close—to their exposed
homes and barns specifically
to deflect lightning bolts
from those vulnerable
structures.

How Not to Get Zapped
Observe the following rules, and teach them to anyone who will

listen. They could save a life.

The first commandment, if outdoors, is to take shelter the moment you hear the rumble of thunder. Lightning can strike up to 16km/10mi from its source. And since light travels so fast, the storm may be closer than you think. Using the flash-to-boom method, a count of ten seconds means the storm is 10 x 330 metres or about three kilometres (less than two miles) away. If your count is anything under 30 (10km/6mi), head for cover.

The emerging science of lightning trauma is called *kerauno medicine*.

Although no place is absolutely safe, some places are safer than others. The best is a large enclosed space like a barn or warehouse. A car is good too, because its metal body (not its rubber tires) will channel the charge safely to the ground. Roll up the windows just in case, and don't touch any metal parts.

If you can't get under cover, shun open areas and tall structures like telephone and hydro poles. Drop fishing rods and golf clubs. Hands off aluminum boats, golf carts, bicycles, and tent poles, which can conduct a fatal charge to your body.

Indoor untouchables include telephones, appliances, faucets, metal pipes and cable television wire. Don't take a shower, wash your hands, or use electrical gadgets during electrical storms. To prevent damage to appliances, some people unplug them at the first rumble.

Indoors or out, should your skin start to tingle or the hair on your arms start to rise, you're in danger of being struck. Immediately move at least 4m/13ft away from anyone else, squat on the balls of your feet (to minimize ground contact), lower your head between your knees (to minimize the target) and place your hands over your ears (to minimize damage to eardrums).

Finally, don't relax too soon. Wait half an hour after the last rumble or flash.

In the event of a strike, apply first aid to the victim: do a quick check for injuries, administering CPR if necessary (there's no danger of receiving a residual shock from the victim); call for help; and move the victim to a safer area while you wait.

Special

Effects

In 1999 *Star Wars* fans flocked to theatres to see the first film in George Lucas's second space trilogy, the long-awaited "prequel" or backgrounder to the first three, released 22 years earlier. They found the special effects even more dazzling. The dinky carpenter shop that had fabricated the original's space ships, planets, and monster suits had gone digital. Lucas realized early on that computers can do special effects better and faster (but not cheaper, at least not yet) than carpenters and sculptors. From that realization had come Industrial Light and Magic, the world's premier digital imaging facility.

I.L.M. and its computer graphics wizards gave us the dinosaurs of *Jurassic Park*, the cyborg in *Terminator 2*, and the Kennedy cameo in *Forrest Gump*. Films like *Back to the Future*, *E.T.*, and *Ghostbusters* also owed a debt to Lucas. Understandably, top-flight filmmakers flock to this facility at Skywalker Ranch in the hills of West Marin, California like pilgrims to a shrine.

Dazzling as digital graphics and special effects are, they are merely lights playing across a cathode ray tube or theatre screen. Not only that, but most of the dazzle happens inside our skulls. Impressions of vast space and breakneck speed are optical illusions.

Hast thou entered into the treasures of the snow? or hast thou seen the treasures of the hail?
—Book of Job 38:22 (NEB).

Compared to a real sunrise or rainbow, this is like watching the moon in a rain barrel. The real special effects extravaganzas occur in the real sky, that global ocean of air in which we all swim. Inconceivably vast and ever-changing, they are remastered every day and free for all to see. Yet except for everyday effects such as the odd sunset or thunderstorm, the northern lights or a spectacular rainbow, not many of us have witnessed those extravaganzas. Some have been lucky enough to see a ring around the sun or moon, maybe even a set of sun dogs or a moon pillar. A few may have spotted iridescent or opalescent clouds, and what the Germans call the *Heiligenschein*. Rare indeed is the person who has witnessed the "Green Ray," "The Glory," or a trio of magnificent crosses shining in the sky.

Some of these phenomena aren't as rare as one might think. It's just that hardly anyone knows when or where to look.

"Sunrise, Sunset..."

When the Russian milkman Tevya sang this poignant song in the musical "Fiddler on the Roof," he was musing about the inexorable passage of time, about how swiftly one's children grow up and leave home, about how old age creeps up on a person. In a very real sense, sunrises and sunsets mark off the days and nights of our lives.

No one is counting, but now and then the spectacle is so grand it stops us in our tracks, making us yell for someone to share it with, or to bring a camera. "Sunrises always work," remarks Joy Gresham to C.S. Lewis (played by Debra Winger and Anthony Hopkins) in the movie *Shadowlands*. So do sunsets, except they are tinged with the sadness of closure.

Sunrises tend to be less dramatic than sunsets, partly because overnight the clouds have cooled, flattened, and descended. On the other hand, it's a proven fact that horizontal lines are more calming than angular ones, which is why morning clouds have that serene and hopeful look—unless a storm is brewing. And morning colour, all shell pinks and pale lemon, tends toward greater delicacy.

Sunsets, on the other hand, tend to be flamboyant, even lurid. That's because the low sun is often shining through cumulus clouds with stratus clouds behind and above, and because the air is cluttered with the day's dust, smoke, and pollutants.

The thing that makes sunrises and sunsets so memorable, of course, is their colour. The palette ranges from thundering reds through vibrant yellows to sombre purples. The preponderance of reds and yellows is due to dust, water droplets, and industrial or volcanic particles in the air. Sunlight itself is pure white, which as Newton proved consists of a spectrum of seven main colours. But the microscopic particles scatter the high-energy violet, blue, and green reflections, letting the low-energy warm hues shine through. (The same thing happens when we look at a glass of skim milk. Hold it away from the light and the milk looks bluish due to scattering and reflection by suspended calcium particles; this is also what makes the sky look blue. Hold it against the light and the milk looks pinkish due to the low-frequency red hues.)

This rosy effect is magnified at sunset and sunrise, when the sun is shining through a greater thickness of air—and the dirtiest layer at that.

Thoreau once remarked that if the stars shone once every thousand years, people worldwide would rush outside to see them.

In a lifetime of seventy years, with clear skies and a good view every day, a person could theoretically watch 25,550 pairs of sunrises and sunsets.

Paradox: We gaze east at sunrises and west at sunsets—yet in relation to us the sun never moves. How come? (Hint: Pretend you're on a train circling a mountain every 24 hours.)

Rainbows & Related Phenomena

Anyone can make a rainbow. All you need is a garden hose with a spray nozzle on a sunny day. Stand with your back to the sun, turn on the water, adjust the nozzle to a mist, and aim it toward your shadow at chest level. If the spray droplets are fine enough, you'll see an arc of colour that moves when the nozzle moves and that disappears when you turn toward the sun. Children know all about it.

Sometimes such a rainbow appears beside a public fountain. A backlit ocean wave can also create a rainbow as it crashes against a cliff. Waterfalls, if backlit and tall enough, are commonly adorned by an arc of coloured light. North America's most famous example is the one that sometimes graces Niagara Falls, with the tour boat *Maid of the Mist* at its foot. Splendid as that rainbow is, I'd just as soon see the one over 9m/30ft Moose River Falls on Nova Scotia's Parrsboro Shore. It's the perfect end to a lovely hike.

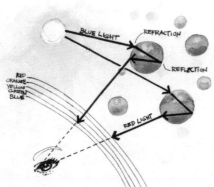

Natural rainbows occur whenever one is looking toward mist-laden air with a low sun behind. This usually happens when the sun breaks through low clouds behind a retreating rainstorm, showing a coloured arch against a backdrop of darker clouds.

What all these rainbows have in common is reflected light. When sunlight passes through water droplets of a small enough diameter, it is refracted (bent) on entering and again on leaving. This splits the white light into two or more colours. Still, we might not see this coloured light were it not also reflected our way by the droplet's concave rear surface. Anyone facing the droplet sees that tiny flash of coloured light and any nearby flashes included in their cone of vision.

In his book *The Nature of Light and Colour in the Open Air*, Dutch professor M. Minnaert faulted the great baroque artist Peter Paul Rubens for getting the sequence of his rainbow colours wrong. He excused Rubens by saying that the artist had likely never really observed a rainbow. It is quite possible; few seventeenth-century artists worked directly from nature.

The combined effect of these millions of suspended spherical prisms is a narrow arc of prismatic colours, arranged in a set order from red on the outside to violet on the inside. Prismatic colours always follow the same sequence—red, orange, yellow, green, blue, indigo, violet. Many people remember the sequence by using the mnemonic device "Run O You Great Big Irish Villain"—but that's hardly fair to the Irish.

Was the English scientist Isaac Newton thinking of rainbows when he proved that white light is really a blend of several colours? It may be that his revolutionary idea came from watching the play of light on a chandelier or on a cut-glass carafe. To test it he chose a wedge of clear glass. Placing the prism in the path of a pinhole beam of sunlight darting across a darkened room, he caught the beam and projected it onto a white screen. He was rewarded by a brilliant fan of coloured light in a continuum from deepest violet to vibrant red. He had replicated the rainbow. That wasn't enough for him, however. By directing his "spectrum" through a second prism, he restored the dot of white light. He also noted that the reds came from the thick end of the prism (like the hues of sunset), while the blues came from the thin end (like the colour of a blue sky).

Have you noticed that the sequence of hues from horizon to zenith in a cloudless evening sky after sunset follows a rainbow sequence? Beginning with red or pink at the horizon, they grade delicately up through orange and yellow into turquoise and deep blue-violet directly overhead. The same thing happens at sunrise. The sky is a rainbow dome.

Airline pilots and mountaineers sometimes see a circular rainbow. This phenomenon appears on the clouds below, centred on the shadow of the plane or observer. The only time a sea-level viewer sees even a half-circle is at sunset or sunrise. The rest of the time the arc is smaller, or partial with one end touching the ground. With the full 180° arch there's a good chance of seeing a second bow outside the primary one, and sometimes a third on the inside, both fainter than the main bow.

The Greeks linked rainbows to the goddess Iris—hence our name for the coloured part of the eye—and saw them as bad omens. Medieval Europeans and Britons fancied the base of a rainbow marked where a pot of fairy gold lay buried—the fabled "pot of gold at the end of the rainbow." The Irish leprechauns supposedly hoarded gold coins underground to buy their freedom if they ever got caught. One can imagine how the myth might have begun. Red and yellow hues dominate in rainbows. At full strength they resemble an arch of glittering gold.

Too bad, though, that when one walks toward a rainbow it always retreats at exactly the same rate. It can seem near enough to touch—yet tantalizingly out of reach. The only time we have a hope is when the bow appears against a shadowed cliff or a dark ridge of conifers. But there's no pot of gold. I've looked. The treasure is in the sky.

I will never forget one evening when my wife and I were driving east between Sackville and Amherst on the Tantramar Marsh. The afternoon had been showery, and now, near sunset, the low golden sun was breaking through. Ahead of us towered a bank of slate blue cumulus clouds, streaked with rain and none too inviting. Suddenly, as we rounded the ridge at Aulac,

Rainbows are actually circular, but earthbound observers see only the part above the horizon, always with a 42° radius. If a second bow appears, it will be larger (50°), with its colours reversed.

we saw a glorious multiple rainbow. Everything in the foreground—hayland, trees, houses, fences—was awash in golden light; everything in the background was lavender and blue, and there, arching over everything, was this 180° apparition. It lasted over half an hour, retreating steadily before us into Nova Scotia almost to Oxford. I thought of the Biblical account in Genesis 9, where God makes this promise to Noah:

I do set my bow in the cloud, and it shall be a token of a covenant between me and the earth....[that] the waters shall no more become a flood to destroy all flesh.

Snow & Frost

Why do snow and frost look white, when in fact the ice they are made of is as transparent as glass?

To answer that we need to look at the structure of both, and at the nature of sunlight. Both snow and frost are composed of ice crystals formed from water droplets, the first while suspended in freezing air, the second while lying on a freezing surface. Although a cube of ice is nearly as transparent as glass, its crystals are arranged differently.

In 1931 W.A. Bentley and W.J. Humphreys published their ground-breaking book *Snow Crystals*. It contained magnified photographic images of thousands of individual snowflakes, no two alike.

Snowflakes aren't the same as snow crystals. Snow crystals are fragments of ice only a few millimetres across. Those big feathery flakes you see drifting lazily down in mild weather consist of numerous tiny crystals clumped together. To observe an individual crystal it's necessary to go outside when fine snow is falling and catch one on some dark heavy material like a woolen glove or a coat sleeve. Hold it close and examine it, being careful not to melt it with your breath.

The most striking thing about snow crystals is that they are flat, thin, and nearly always six-sided or six-pointed (hexagonal), like the ones on Christmas cards. All snow crystals obey this Rule of Six.

Since individual snowflakes come to rest at different angles, the reflected sunrays bounce back at different angles with enough refraction to split the colours.

Being flat, they reflect most of the light falling on them. As Newton proved, sunlight is a blend of red, yellow, green, and blue light, plus their combinations giving the optical sensation of white. (Light has four primaries, while pigments have three.)

I often see the process in slow motion, as it were, as I walk past my study window where a crystal hangs. When the sun is over my shoulder, I see a series of rainbow-coloured flashes as I pass through the fan of reflected light.

This is the effect we see, multiplied a million-fold, when the sun shines on a field of fresh snow on a cold morning. Because the crystals are as yet undamaged, they reflect the sunlight upward in all directions like the facets of a diamond, and each flash reaches our retina with one of seven colours. Come back in a few hours, however, and the snow may look white. By then the individual flakes may have melted and clumped, or a breeze may have tumbled and rounded them. Now the snow reflects all the colours indiscriminately, producing white or grey. Aluminum paint gives a similar effect.

Outdoor frost, though beautiful up close, is seldom as reflective as snow. Since it is formed by the passage of moist air over a cold surface, surface texture and moisture govern its thickness and patterning. Usually it's only a few millimetres thick, with a pebbly or crystalline pattern. When multiple layers accumulate in a thick coating over trees, wires, etc., we call it hoar (white) frost. The cause is a gentle flow of mild, moist air after intense cold. I've seen a related type of frost up to 4cm/1.5in deep on damp, rotting wood in the forest.

My favourite frost formation occurs on drafty farmhouse windows covered by old-fashioned storm windows that leak a little when it rains. Our house has several such windows. As water vapour swirls between the outer and inner windows, it gradually builds a field of crystals upward on each inside pane—feathery plumes at first, then an increasingly interlocked matrix. The last parts of the glass to be covered are the top and those inner surfaces exposed to leakage of cold outside air.

The resulting works of art—delicately pebbled fronds and tendrils and arabesques of clear and cloudy ice—are richest after very cold nights, when the outer panes accumulate a couple of millimetres of ice. The best effect is at sunrise. As the orange globe lifts above the horizon, the backlit panels suddenly blaze with silver and gold with pink, green, and blue flecks. By moonlight it is like gazing into a frozen pewter jungle. As the room heats up, these glories fade. Modern multi-glazed windows rarely exhibit such wonders.

Haloes, Sun Dogs, Pillars, Glories

All of us have seen sunbeams, those bright rays flaring out from behind dark clouds or hills in searchlight patterns. They're caused by water vapour and dust scattering the light. And we've all seen man-made jet contrails, high-altitude linear clouds that drift sideways in wavy lines as the wind pushes them, gradually widening until they vanish. They consist of frozen water droplets from the plane's exhaust. The moister the upper air, the longer they last.

But few seem to notice the less obvious signatures of high-altitude ice crystals. When I point one out, people usually express

Fresh snow and ice is so reflective that it sends over ninety percent of the light that falls on it back into space. If it wasn't, the Greenland and Antarctica ice caps would melt faster than they do. Scientists call a planet's reflectivity its *albedo*.

surprise that such things exist. Which is a pity, for these high-level spots, patches, and rings are uniformly lovely—and useful in forecasting weather. They are caused by light passing through and/or bouncing off tiny ice crystals. Like snow crystals they are six-sided, but come in a wider range of shapes, including columns, platters and bullets, producing peculiar effects.

Haloes

The lower the cloud deck, the larger the circle. That's why oldtimers say the larger the halo, the sooner the weather will change.

These are partial circles that appear at an angle of 22 or 46 degrees around the sun or moon due to light refraction (bending) and reflection (mirroring) from high-altitude clouds of oval ice crystals. They usually signal the arrival of a warm front. As the incoming lighter air rides up over the cooler air below, it forms a gradually thickening wedge that, at a certain depth, produces the halo effect. When the cloud cover is complete we get a full ring.

Something similar occurs when a wet tree is backlit by a street-light or the moon at night. Since only the twigs in our cone of vision reflect any light, we see concentric arcs of light which, in dense branches, almost form a ring.

By day the halo is a faint, ashy yellow edged with pale blue; by moonlight it is a paler version of the same.

Sun Dogs

How would you like to have more than one sun in the sky? George Lucas played with this idea in *Star Wars*. It certainly gave his alien planets an other-worldly look. We earthlings must make do with one sun, yet sometimes even sober people see two or three.

Sun dogs (also called mock suns, sun hounds, or parhelia) occur ten or twelve times a year in clear, cool weather, often near sunset. They are caused by a thin cloud of platter-like ice crystals veiling the sun. The best times to see them are during the morning or evening in late winter and early spring. The ice cloud, by refracting and reflecting the light, creates a multi-coloured oval spot 22 degrees out on one or both sides of the sun. Sometimes a partial ring is present too. These displays last an hour or so.

Like the halo, sun dogs often foretell a shift to milder weather, meaning rain in summer and snow in winter. The colours are pastel pink, yellow, green, and sometimes blue. Ancient people believed sun hounds foretold troubled times.

A variation of the sun dog is the pillar, a vertical shaft of faintly coloured light above or sometimes below the sun. Pillars are caused by light bouncing off the undersides of flat, tilted ice crystals suspended in still air at high altitudes. On rare occasions the moon sports a pillar.

I saw one of these on a bitterly cold, windless, February night. As I walked our dog, thinking it might be too cold for her, I noticed the full moon rising out of a cloud bank. Above it a golden pillar shot straight up like a searchlight, with a fainter parallel pillar on either side. The colour, seen against night-grey clouds, glowed like a Rembrandt portrait.

SUN PILLAR

A ground-level variation of these ice clouds is "diamond dust"—frosty, windless air filled with darting scintillas of white light like angelic mosquitoes.

Sometimes ice crystals create a ring of coloured light around the sun or moon called a corona. (Strictly speaking, the corona is the ring of fire seen during a total solar eclipse.)

The Heiligenschein

German for "bright light," this happens when a person looking at his or her shadow on close-cropped, dew-laden grass sees a spot of bright light around the head and shoulders. It is best seen looking down from a hill or ridge with a low sun directly behind. I've seen it while walking the Fundy dykes on a clear evening in the fall. At times the halo is vivid enough to photograph. It usually startles me at first, but then I get to like the illusion of sainthood. The colour is misty-silvery with a hint of gold. Persons walking together will each see the halo around their own head.

The Glory

This is the same thing seen against a wall of mist, with richer colour and more haloes. On one occasion a five-fold glory was seen around a person's head. Professor M. Mennaert shows a spectacular one in his book *The Nature of Light & Colour in the Open Air*. Unfortunately it's in black and white. The effect is very like the traditional haloed Madonna image. As with the *Heiligenschein*, the angling of refracted light is so exact that, while individuals in a group of people can each see their own shadow, the glory will appear around their own head only.

An even rarer apparition is that of one or more crosses in the daytime sky. This combines twin sun dogs bridged by a horizontal line of light and bisected by a sun pillar. The pillars may or may not be tilted off the vertical. Sky crosses, because of their rarity and religious overtones, excite superstitious people and get reported in the popular press, especially if they coincide with a disaster of some kind. Four women felt privileged to see one over Debert, NS around 1992.

Green Rays

Also called the "Green Flash," this rare and beautiful effect occurs precisely at sunrise or sunset, and lasts barely one second. It is caused by sunlight being scattered by ice crystals in such a way that a momentary vertical flash of emerald green appears against the clear sky.

Northern Lights (Aurora borealis)

One winter evening after supper when I was twelve or so, a friend banged on our door and hollered for me to come see something. Outdoors, the whole northern sky flickered with swaying curtains of rainbow hues that outshone the stars behind. Next day the old retired seaman next door frightened us by announcing that they meant the end of the world.

The aurora borealis or northern lights (New Zealanders call them aurora australis or southern lights), like other special sky effects, owe their beauty to the sun, but for different reasons. As their neon-like appearance suggests, they are electrical in nature. They occur when a burst of charged particles from the sun collides with air molecules over the poles. Channelled and accelerated by the earth's magnetic field, they result in fluttering curtains of pastel rose, lavender, green, and gold. Auroras are most common in autumn and early winter, and are linked with cyclic sun spots and solar flares. Some say they can hear the lights rustle like silk. It's a tantalizing thought, but this has never been proven.

20

Backyard

Forecasting

One summer several years ago, a young man returned from Toronto to the Maritimes seeking his rural roots. As he entered the village store, he passed a benchful of oldtimers yarning and whittling. Not wishing to seem unfriendly, he remarked, "Looks like rain."

The oldtimers eyed him down and up but made no reply. Instead, one of them asked his name. The young man obliged.

"Then you'd be the grandson of...?" inquired another.

"That's right," said the young man, feeling better.

"And you say it might rain? Well, I daresay it just might...."

There was a time when one could be burned at the stake for publicly forecasting the weather. Britain had such a law on its books as late as 1960. It's fortunate Canada has no such law, as I've been doing it for years. Well, not in public, but at home and among friends. You can still get into trouble for it, though. One cloudy day at a local church picnic, forgetting my place as a newcomer in those parts, I remarked to a local farmer that the sky seemed to be clearing. The farmer regarded me for a moment, looked at the sky, and drawled, "So you're forecasting the weather, are you?" What he was telling me, I think, was that would-be forecasters must earn the right to practice under someone else's sky.

Fortunately, weather belongs to us all. The storm that drenches southern New Brunswick this morning may drench Nova Scotia and PEI tonight and sail on to Newfoundland tomorrow. Weather is bigger than us all.

I've often wondered why we talk weather so much. Few of us work outdoors any more. Most of us don't have to decide whether to hay tomorrow, or whether our woodlot road is frozen hard enough to haul logs over, or whether the northern drift ice will let us set our lobster pots next Monday. On the contrary, most of us live and work and shop under artificial light in indoor spaces with controlled temperature and humidity. Most days our biggest weather concerns are what to wear, when to mow the lawn, whether to chance driving to work. All we really need to know is whether a storm will disrupt our plans.

Early last February, our youngest daughter was flying in from Vancouver via Toronto. I'd agreed to meet her at the airport at 9:35 P.M. and drive her to Halifax. A blizzard warning was in effect, but the forecast said it wasn't due until "overnight." Outside, the sky was calm after a grey afternoon.

Some are weatherwise, some are otherwise.

—Benjamin Franklin

Entering the airport at 9:15, I heard the announcement that Joyce's flight had been re-routed at Toronto and that its passengers would now arrive at 12:15 A.M. Outside, the airport flags were snapping loudly in a rising east wind, yet no snow was falling. I went back in to wait. Soon the wickets were closed and the cleaning people started their rounds. Wishing I'd brought a book, I cruised the empty terminal looking for a store still open. The gift shops and restaurant were all closed and barred. Two young men riding scrub machines with blinking coloured lights made their noisy circuits like lawn mowers in tandem.

At 11:30 P.M. I went out again to survey the sky. Now the flags were snapping harder, their halyards beating time against the metal poles. A fine haze of snow dimmed the car park's mercury lights.

Hmm. If the flight arrived on time, we could easily drive the short distance to Halifax before the blizzard came. I'd wait it out there overnight and return to Truro the next day. But if she came even an hour later, or the flight was cancelled, we wouldn't be going anywhere.

I booked a double room at the nearby Airport Hotel for $119 plus tax and resumed my vigil. Overhead, through the glass ceiling pyramid that gave us a dramatic view of the whirling airport beacon, streamers of snow were already dimming the beam.

Around midnight the PA crackled to say the flight was cancelled and that the passengers would arrive via another airline the next morning at 10:50. I buttoned my jacket, pulled on a toque, and stepped out into the blizzard. Already my car was half covered. White-outs on the short drive across the overpass to the hotel convinced me I'd made the right decision. My daughter arrived on schedule the next morning, travel-weary but okay.

Being on tenterhooks about the weather is less a problem for today's Maritimers than it was for, say, a nineteenth-century schooner captain. Yet deep down we still harbour a sort of residual weather anxiety. I think this may be a survival instinct hard-wired into our brains like the fight-or-flight instinct. After all, humans have grappled with the elements since prehistoric times.

When our Cro-Magnon ancestors were laboriously decorating the caves of Lascaux and Altamira with ochre and candle soot over twelve thousand years ago, their world was still in the grip of the last Ice Age. And until recently our rural ancestors anxiously monitored sunrises and sunsets, clouds and stars for

I was born with a chronic anxiety about the weather.

—John Burroughs, *Is It Going to Rain?*, 1877

storm omens. Even now most people instinctively glance at the sky as soon as they wake in the morning. Something in us needs to know.

The media, ever swift to seize a market, have obliged. Several times a day—24 hours a day on the weather channel—telegenic people with pleasant voices spoon out weather information. Reading teleprompter cards, pointing vaguely at maps that aren't there, they tell us what's coming and why.

We don't seem to mind that few of them actually compile the forecasts. We know they're merely presenters. Yet we trust them. Radio tries to do the same and, perhaps because radio has been doing it longer, somehow comes across as more reliable, though they use the same Environment Canada data.

You may ask why we, deluged as we are by packaged weather, should do our own forecasting. I can think of at least four reasons.

First, most forecasts are put together soon after midnight for release at 5:00 A.M. By the time we see or hear them, they're already two or three hours old. A lot can happen in the interval. Second, every forecast is a blend of data from several stations. It never applies exactly to where you live. Third, government cutbacks in the late 1990s may have compromised our weather services.

These reasons apply especially to hurricanes, which are famously unpredictable, and to thunderstorms and tornadoes, which are usually homegrown. Granted, modern forecasters, thanks to satellites, computers, and instant communications, get it right more often than they used to. But East Coast weather is still notoriously hard to predict.

The fourth reason for being your own weatherperson is that it will get you looking at the sky again. Would you believe that very few us ever look up unless something spectacular like a thunderstorm is going on? Jack Borden, a former Boston TV news reporter, proved this one dazzling June day by interviewing people on the street for the evening news. His question was, "Please describe the sky right now without looking up." To his utter amazement, not one person could.

"Oh," we say, "that's big city life for you; too many tall buildings, too much distraction; he should have interviewed some country people." Perhaps; but it's safe to say none of us pay much heed to what's happening overhead. So if, as Ralph Waldo Emerson put it, the sky is the daily bread of the eye, then we are visually malnourished.

In February 2001, East Coast mariners and others accused Environment Canada of issuing inaccurate forecasts. The forecasters agreed that in the previous five years their accuracy had declined from 75% to 60%. They blamed staff reductions and aging equipment due to budget cuts.

They say certain "primitive" peoples could predict local weather correctly about eighty percent of the time. To do so they sniffed the wind, watched sunrises and sunsets, noted rings around sun and moon, observed the movements of wildlife. For them, forecasting was a crucial survival skill. How else could a chief decide when it was safe to send his hunters on long winter journeys or dangerous summer whaling expeditions?

Our forebears were no slouches at weather forecasting either. Using homemade instruments and their own traditional lore, the best of them scored almost as high as the natives did. How else could a farmer decide when to cut his hay? A schooner captain when to hoist sail on a risky voyage? A sealer when to venture out of sight of the mother ship? A fisherman when to head for land? How could a logger predict the spring breakup so as to hire river drivers to bring down the winter's cut? Although most early predictions were for a day or two at most, it took a lot of savvy in that unwired world.

Tradition states that Nova Scotia's Mi'Kmaq sometimes canoed 100km/60mi from northern Cape Breton to southwestern Newfoundland to trap or hunt for a season, a feat of navigation that demanded skillful forecasting.

Of course it helped that one's life or living rode on the outcome. Still, they made mistakes. Experienced schoonermen, fishers and sealers perished in unexpected hurricanes and blizzards. Nova Scotia's Sable Island isn't nicknamed "The Graveyard of the Atlantic" for nothing.

Fortunately, your backyard forecasting will seldom, if ever, be a life-and-death matter. Hikers needn't lug a barometer, max-min thermometer (to test humidity), cloud charts, and weather tables. Common sense and good observation should be enough. At home, the same skills plus a few simple instruments are all one needs.

Knowledge of local conditions is useful too. For instance, here by the inner Bay of Fundy a westerly wind usually precedes the incoming tide by a half-hour or so, especially on the new moon and full moon. This often clouds the sky for an hour or so. In winter the same breeze can bring short-lived snow flurries that a stranger might mistake for a real storm.

Knowing such things, and with the help of a few simple instruments and good records, you should be able to make decent one- or two-day forecasts. But take my advice; keep your hobby to yourself for at least a year.

Serious weatherwatchers can't do much without an outdoor thermometer, a weathervane, and a barometer.

Thermometer

Your thermometer can be an alcohol or a mercury model; it doesn't much matter, so long as the markings are big and easy to read even through a skim of snow or fog. It helps if the scale is inscribed on the glass and not part of the frame, as the two sometimes get jiggled or rusted out of line. It also helps if the scale has both Celsius and Fahrenheit notations. You don't want to be constantly translating.

Choose a spot with good air circulation, preferably on the shaded (north-facing) side of your home. Don't put it near an air conditioner or a dryer vent that might elevate the readings. Mount the thermometer at about waist height where the wind can't shake it and where excess water can't drip on it and freeze. (Note: The max/min thermometer is similar but allows one to record warmest and coolest temperature in any one period.)

The word *vane* comes from the Anglo-Saxon *fane*, which means flag.

Weathervane

This is essential because winds usually shift before a change. You can make one at home from wood or metal. It needn't be fancy, but it must respond to light winds and yet withstand a gale. Early weathervanes used a "wind-flag," a strip of cloth nailed to a pole. Airport windsocks and Japanese sock-kites use the same idea.

EARLY
WEATHERVANE:
PINE BOARD

A better rig is a wooden or metal arrow balanced on and fastened to a metal spindle that turns on a hard surface that won't soon wear out. Mine is clamped to a long vertical spike inserted down through a drilled block of wood nailed to the top of our clothesline pole. The spike is seated on a square of window glass which lasts indefinitely and never needs lubrication. Make sure the tail has more drag than the head, or your vane will swing sideways to the wind. Wooden parts last three or four years, twice that if treated and painted; the spike will last ten or fifteen years (longer if greased); and the glass even longer if ice buildup doesn't crack it. I've replaced my vane only twice in twenty-five years.

OLDTIME WEATHER VANE

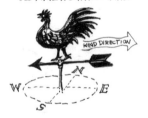

You can also buy a fancy weathervane, or have one built. Some people surmount the vane with a rod carrying the four compass points properly oriented. Secretly, I've always wanted a metal rooster vane with lots of clashing tail feathers, all made of copper so it will age to that lovely jade green patina old statues wear.

My artist friend, David Blackwood, adorned his Newfoundland studio with a black metal sperm whale. When my red Douglas-fir plywood arrow fell apart last winter, I replaced it with a silver-painted Atlantic salmon.

Using a rooster image on weathervanes dates back to Pope Nicholas I of the ninth century. He decreed that Russia's church spires be topped with such a vane to remind parishioners of Peter's denial of Jesus the night before his crucifixion.

Barometer

Oldtime farmers used to hang a length of manila rope from a barn beam to warn them of a change in weather. It worked, too. The rope would unwind before wet weather and rewind before dry weather. A pointer suspended over a circle marked on the floor showed them which, and how fast the change was coming. Though it actually registered airborne water vapour, not atmospheric pressure, it functioned as a crude barometer, since low pressure systems usually bring rain and vice versa.

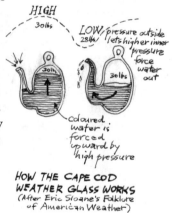

Meteorologists use a mercury model (a closed-off glass tube inserted open end down in a pool of mercury). An early version of this was the American "weatherglass." This egg-shaped glass sphere had a spout like a teapot (but no handle; it was hung on a wall), and contained air and water.

When outside air pressure was low, the trapped air inside the glass expanded downward, pushing water up the spout until it poured out. "When the glass spills over," said oldtime Cape Codders, "so will the clouds in a little while." It was from this device, and not from the mercury or aneroid barometer, that we got such expressions as "the glass is rising."

The glass is falling hour by hour,
the glass will fall forever,
But if you break the bloody glass
You won't hold up the weather.

—Louis MacNeice, *Bagpipe Music* (Undated)

HOW THE CAPE COD
WEATHER GLASS WORKS
(After Eric Sloane's Folklore of American Weather)

The small aneroid model is a lot cheaper and almost as good as the mercury model; indeed, aircraft altimeters use the aneroid principle. The device has two hands, one to show actual pressure and the other to record movement over time. An aneroid barometer fitted with a rotating graph-covered drum and a pen can trace changes in pressure over days and weeks; such tracings are called barographs.

The clock type, encased in mahogany and brass, looks quite handsome on a wall. During the 1800s clock-type circular aneroid barometers became a popular decoration in English homes, but the fad never really caught on in America.

Ideally the barometer should be adjusted to sea level, meaning 1013.2 millibars of mercury or 29.92 inches of mercury. But so long as readings are comparative (has the pressure risen or dropped?), it doesn't really matter. For years I used my aneroid instrument without adjusting it; the thought intimidated me. Then one day I noticed the little adjusting screw at the back.

CLOCK BAROMETER

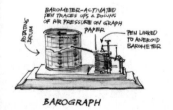

BAROGRAPH

Taking the barometer down to the shore, I set it to 29.92 with my pocket knife, and hung it up again. The range from high to low pressure at ordinary elevations, barely 950-1030 millibars (28-30.4 inches), is hardly noticeable. But at higher elevations adjustment is important.

These three items—thermometer, weathervane, and barometer—will give amateur skywatchers all they need for basic forecasting. In time you may want to invest in a cup anemometer (for clocking wind speeds), a hygrometer (for measuring humidity), a sunshine gauge (a mounted glass sphere called a Campbell-Stokes Sunshine Recorder, which focuses the sun's rays and traces a daily scorch mark across a piece of cardboard—a delightful toy), and a rain gauge.

Setting Up to Forecast

Okay, let's say you now have some notion of how weather works; how the sun heats the land and the sea unequally, creating balloons of air that rise and are nudged northeastward by the earth's rotation and the Coriolis Force; how cold dry air collides with warm moist air masses from west or south of us, wringing out precipitation and creating wind; and how, when the westerlies finally dominate, the skies clear.

Likewise you understand how weather is continually recycled; how the storms or calms in Baltimore or Toronto are tugged northeast by the earth's spin and become our storms or calms a few days later, modified in passage yet still recognizable; how

Probably the last completely accurate forecast was when God told Noah there was a one hundred percent chance of precipitation.

—Internet humour

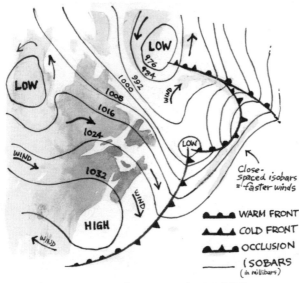

FRONTS, HIGHS/LOWS & ISOBARS

Maritime weather becomes the weather of Newfoundland, the outer Grand Banks, and perhaps even of Iceland and Ireland.

Finally—and this should comfort the home weatherwatcher—how these processions of opposing air masses form and re-form over our heads with almost clocklike regularity like dancers in a minuet, differing in details but ruled by the same laws...making three-day forecasts feasible.

We're ready. Reminding ourselves that even the experts can't predict accurately beyond three days, we launch out. Our modest goal is to predict what the sky will do in the next 24 to 36 hours and be right at least half the time without using outside help. We don't want to be like the local weather prophet who, asked about tomorrow's weather, said he didn't know because his radio had quit. We want to be able to observe the signs, consult a few instruments, and come up with a prognosis.

Sample Observations

The key to success is to skywatch every day, jotting down what you observe. Look at cloud forms, sky colours, wind direction and speed, shifts in dampness and temperature. Ask yourself questions like these:

What colour was the sky at sunset last night? At sunrise this morning?

In the last day or two, have you seen a ring around the sun or moon, or rainbow spots beside the sun?

Is there dew or frost on the grass?

Is the air warm or cool/cold?

If there is wind, what direction is it blowing from, and at what approximate velocity?

Is the sky clear or cloudy?

If cloudy, are the clouds lumpy, flat, or a uniform grey?

If there are clouds at several levels, from what direction are the highest ones coming?

Is the barometer rising or falling, and if so, how fast?

WARNING: Never stare straight at the sun, even when wearing

sunglasses or using a dark filter. To do so can damage your eyes and even cause blindness. It's okay to glance directly at the setting or rising sun.

What's Normal Anyway?

Because it helps to know what is typical weather for our part of Canada season by season, here's a brief review.

Spring is generally a drawn-out affair with several relapses. As Garrison Keillor put it, "March is the month God created to show people who don't drink what a hangover is like." Gardeners expect the last hard frost around the full moon in late June. (Officially, June 21 is midsummer, but in coastal regions it takes another month for the ocean to lose its chill.)

For the same reason summer is also tardy. It is usually late July or early August before anyone complains of the heat. As a result, our autumns are quite warm, a time of misty mornings and toasty afternoons. Hard frost is rare before late October except at high elevations or in Labrador. The ocean, now as warm as it will get, acts like a hot water bottle in a cold bed. This is obvious near the Bay of Fundy. If a frost is forecast, we don't cover the tomatoes unless the tide will be out between 3:00 and 4:00 A.M. If the tide will be in, it will ward off the chill.

September is hurricane month all along the Eastern Seaboard. We commonly get the tail ends of several, one or two of which may be severe, especially along Nova Scotia's southern shore and in the Bay of Fundy. These whirlpools of warm moist air generally fizzle once they hit the chill Labrador Current.

Around mid-October we usually get ten days to two weeks of warm weather called Indian Summer. No one knows for sure where the name came from, but it probably originated south of the border where agricultural First Nations people would be harvesting their maize, squash, and beans about that time.

Our first tentative snowfall can be expected in late November, sometimes earlier in New Brunswick's western uplands and in northern Cape Breton. Mainland Nova Scotia's first serious snowfall usually arrives in late December, often after several false starts—the influence of that ocean warmth again. Winter finally sets in around the turn of the year.

Even then, temperatures rarely stay below -10°C/13°F for more than a few days. Now we get a lot of sleet and freezing rain, and roads grow treacherous with black ice. For better traction, smart people put extra weight in the rear of their vehicles—bags of sand or hen grit in cars, a load of firewood or cement blocks in pickups. And so the revolving year brings us back to spring.

The Beaufort Wind Scale

How do we describe various velocities of wind with no way to clock them? Sir Francis Beaufort of the Royal Navy mulled this over too, and in 1805, the same year the British fleet defeated Napoleon's off Trafalgar in Spain, devised such a scale. It told mariners how many sails a full-rigged ship needed to take in under various wind conditions. Thus, a light breeze was Beaufort Force 1 (i.e., reef one sail); a bad storm might be Force 11, and so on. Later his scale was adapted for landlubbers.

Some samples:

Force 0   Wind Speed—1kmh/0.6mph; Wind Type—calm; Effects—smoke rises straight up.

Force 3   Wind Speed—12-19kmh/7-11mph; Wind Type— gentle breeze; Effects—wind extends light flag, leaves and twigs in constant motion.

Force 6   Wind Speed 39-50kmh/23-30mph; Wind Type— strong breeze; Effects—large branches in continuous motion, utility wires whistle.

Force 9   Wind Speed—75-85kmh/45-51mph; Wind Type— strong gale; Effects—branches break, shingles blow from roofs.

Force 12   Wind speed—118+ ; Wind Type—hurricane; Effects—severe and extensive damage.

Note: For the full Beaufort scale, see Resources, pages 212-214.

Some Sample Forecasts

Here are some signs and what to expect. It helps to note barometric pressure several times a day, and preferably to jot it down (or, on the aneroid type, reset the indicator hand). That way one knows whether a low or high is coming, and how severe the winds could be.

A simple way to locate a system is to stand with one's back to the wind and extend one's arms. The left arm points to the low ("L to L"), the right to the high. Since winds in our hemisphere blow counter-clockwise around a low, and clockwise around a high, you'll have a good idea from which direction the winds will come, and how they will shift as the system tracks across our region. A system may take several days to do so.

Clear Dawn after Red Sunset, Cool Air, Wind Westerly, Barometer High (around 30)

The classic fine day scenario, especially if you see dew or frost on grass, car tops and roofs. If there are high-altitude clouds, try to determine their direction of flow. If it's from west to east, expect at least two days of sunshine.

Cloudy Pink Dawn, Easterly Breeze, Air Damp, Barometer Falling

Rain or snow coming within 24 hours, especially if there's no dew or frost and the wind is southerly or easterly. The pink means there's a lot of moisture in the air. Likely a warm front is moving in, so the precipitation may last two or three days.

Sunset Bleary Yellow, Southwesterly Wind, Air Mild, Glass Falling

If it's autumn, good chance of high winds plus rain or snow overnight. If during the storm the winds shift from south to west (clockwise), expect clearing within 24 hours; should they shift counter-clockwise, precipitation may persist two days or more.

Clear Sky, Brisk North to Northwesterly Winds, Unseasonably Cold, Glass High

A slow-moving mass of arctic air, likely to persist two or three days, with wind and occasional showers as warmer air eddies along its edges.

Overcast, Snowing Lightly, East to Northeasterly Winds, Glass Falling Slowly

Good chance of snow for a few days, especially if flakes are small. If winds shift toward the south, snow will turn to freezing rain, then rain, followed by gradual clearing from the west.

Clear Morning, High Fluffy Clouds to South, Air Warming, Glass High but Falling

A warm front moving in. Expect high cirrus or stratus clouds to hide the sun, gradually lowering and darkening with southerly breeze. Expect a gentle, soaking rain in a few hours that may increase and persist for a day or two.

Morning Air Unusually Warm & Humid, Wind Shifting from Southerly to Westerly, Afternoon Cumulus Clouds Rising

People joke about how often it rains on weekends, not realizing there is a connection. Recent evidence proves that factories and commuters generate so much heat and pollution Monday to Friday that we literally create clouds and "seed" them with dust particles that bring on precipitation.

Turning into Dark Thunderheads, Loud Static on AM Radio
Look for a thunderstorm within the hour (take proper precautions).

Light, Warm, Moist Winds Blowing Over Frozen Land or Cold Water (OR Water Much Warmer Than Air), Clear Sky, Warm Rain after Cold Spell
Expect fog for a while.

Scores of other combinations are possible, each with different outcomes. In general, two or three indicators give better accuracy than just one.

Importance of Keeping Records

You can boost your accuracy a lot by recording observations of the symptoms we've been talking about. Make the entries at the same time each day. The best way is to keep a notebook. An office planner with dated pages is okay, but expensive and cluttered with "To Do" stuff. Ring binders tend to lose pages over time. The best choice is a ledger-type journal such as stationery stores sell. Ruled pages are useful, but if you like to sketch cloud shapes, etc., plain pages will serve better. When you finish taking the day's notes, jot down your forecast. That way you'll have an ongoing record of your failures—and successes!

21

Keeping

a Weather

Journal

Off and on over the years—as the fancy took me, or when something unusual or dramatic caught my eye or ear—I've kept notes about the weather. It became a habit, like noting the arrival of a cardinal, or the flowering date of shadbush.

Keeping a weather journal will enrich your life as a skywatcher, guaranteed. It has enriched mine. Writing things down, whether in a journal or in a letter to a friend, imprints them on one's mind. Sketching has the same effect. Rereading those journal notes a month or a year later, we relive the experience. Journal-keeping also becomes a valuable record of local weather from month to month and year to year. And if we include details of daily living, we have a *journal* in the original French meaning— a record of our days.

In my case it happened the other way. Weather news began to infiltrate my regular entries. And when my daughter gave me a sketchbook journal a few Christmases ago, I began drawing in it, unhampered by ruled blue lines. That was better still.

What follows is a sampling of my own journals, arranged by month. While most of these entries describe Nova Scotia weather, they could apply to almost anywhere in the northeast. And although my selections span the twenty years from 1975 to 1995, I've arranged them in twelve consecutive months.

Scanning them now, I see that these jottings, though supposedly concerned with outer weather, were equally concerned with what Robert Frost called inner weather. Outer or inner, weather infuses our whole existence.

The next night, woke to a soft drumming on the roof and a sound of gurgling water.
—Author's journal

January 3
Cold, cold weather for several days, from -10°C by day to -25°C by night; air cuts like a knife. Wood stove no longer able to keep oil furnace from cutting in. Real danger of pipes freezing. Then, high furry clouds sliding over the sun from the south, a relaxing of nature, wind soft on my cheek, fingers no longer scrammed with cold, even without gloves.

The next night, woke to a soft drumming on the roof and a sound of gurgling water. This morning the fields are zebra-striped with tawny gold amid the white. Ditches are noisy with running water; shallow ponds mark the corners of fields. A post-glacial world in miniature.

Ten days of this now, and all traces of snow have vanished except some dirty heaps piled in mall parking lots, and the stumps of snow-people which children have made and loved and lost. Or snowbanks sulking in sunless north-facing corners of fields shaded by trees. Fog rising like steam from those snow-banks, tracks melting from round hollows to ice prints on the ground.

In the woods, sugary old snow is littered with half a winter's catch of fallen twigs, needles, cone scales, catkin bracts, seeds of birch, ruffed grouse feathers from a weasel's winter supper. Mice and vole tunnels lie exposed to the sky, their hard-packed little thoroughfares raised like varicose veins that zigzag across glistening slatey blue ice. Slippery out, too slippery for walking on slopes. Freezing rain has dragged tree branches and telephone wires to earth; cars slide into ditches.

And now the North Wind again, messenger from the Arctic, that world of hard white air, dry as a bone, cold as a stone. Today another massive front arrives, shoving the warm air back with thunder and much weeping until the ground is again rock-hard and the lakes again pewter-smooth. And now more snow.

February 5

Overnight we went from a cobalt blue sky swept by a cutting westerly to a featureless grey sky and absolute calm. But all morning the barometer fell and fell, and toward dark a northeast wind began toying with the loose snow, and a blizzard warning is issued. But at 9:30 P.M. I was to pick up our youngest daughter at the airport, an hour's drive away—unless the Toronto-Halifax part of her trip was cancelled. It wasn't, at least not till later, too late for me, and we both ended up storm-bound, she in Montreal and I in the airport hotel. She arrived from her West Coast visit on another airline at 10:50 A.M. Through fog and wet I chauffeured her home. Some of the snow had melted but she, starved for whiteness, rejoiced in what was left.

March 20

This afternoon, after threatening all day, a late blizzard struck. Being at work and busy with urgent jobs, and not wishing to leave for the fifteen-minute drive home until I had to, I kept checking how the snow in the street was behaving. It's deceptive in town. Wait until things are pretty bad there, and chances are you'll have to abandon your vehicle part way home and walk home—or worse. Had trouble convincing Tony of this. He lives farther away, but he's on the north side, where the wind blows overland and picks up far less snow. That makes all the difference. Besides, our roads are framed by open fields, not woods. They drift up faster, especially where the wind comes off the water.

I was right; I had to abandon the car partway. Driving clockwise around Allison's corner, going slow and craning out the open window because the wipers couldn't keep the windshield clear and everything looked white, I had to steer by the fenceposts to my left. This worked fine until I met the drift that nearly always forms at the turn. The VW Bug, being smooth underneath, tobogganed up onto the drift and hung there, tilted like a sinking ship. Couldn't open my door, and feared climbing out the window lest my weight roll the car on me. So climbed up the passenger seat, rolled down the window, and wriggled up onto the snowbank. But instead of walking home, I took out the shovel and dug away all the snow except two pillars holding up the two starboard wheels. Then, with the flat of the blade, whacked them out from under. VW landed foursquare on its wheels and bounced. Back in business! Cleared the drift, got around the next bend, good traction, making good progress...but the snow was just too deep for so low a vehicle. So I left it well-marked, grabbed briefcase and groceries, and slogged the last quarter mile head-down with the wind on my quarter, thankful things weren't worse.

April 16

On the way home from town, stopped to watch Fundy's rust-red tide pour into the Salmon River at Yuills Island on the full moon. The narrow channel was choked with jostling ice pans trying to get upstream. No ice stays long in one piece on this bay. The older pans, worn oval, were deep-keeled with muddy flanks, and kept plowing into the bottom mud, where they  stuck till the tide floated them free; the newly frozen ones were shallow and angular, and rode over the grounded ones. In bright sunlight, the new snow on each made the water look like churning blood. With the continual roar of grinding ice, the hiss of waves, and the stiff cross-wind that sent pink spray across the pans and me, the effect was one of savage beauty and irresistible force. Yet the high tide never breached the dyked marsh. And the marsh, its prairie of ochre grasses marbled with snow, seemed doubly peaceful.

May 25

Noticing the sky turning grey with low cumulus as I cleared the red currant bushes of winter's tangle of dead grass, I stopped and took the washing off the clothesline. For a moment I leaned against the bench to rest and to survey sky and fields. Then, *phutt!* A hummingbird was hovering among the elderblossoms at eye level. The phrase "ruby-throated" didn't come to me until I saw the squarish scarlet napkin under its beak. The red patch was scarcely larger than my middle fingernail, and, while we stared at each other, it winked out. Suddenly the tiny bird was drab olive green. It perched within two metres of me and flicked out its long tongue like a silver fishing line. Then a car went by, and it flitted away a few metres, presumably to keep an eye on each of us. When the bird turned to whiz away, the ruby spot flashed again. The first drops of rain reminded me where I was.

June 19

Yesterday dawned balmy, moist and windless here on the south side of Cobequid Bay, Nova Scotia. For June it seemed a typical Minas Basin morning of the sort that transmutes to warm sunshine by 10 A.M. But the Cobequid Hills across the bay said otherwise. Their glowing cobalt blue spelled rain. So did the southerly breeze and the plunging barometer. Sure enough, around noon the sky thundered once and celebrated for half an hour with a vertical torrent. Then for another two hours it kept people under cover with sporadic showers.

Around four in the afternoon, our weathervane hitched itself from southwest to west and the barometer inched up. Dry Quebec air began to shred the low, grey cloud deck, revealing swatches of faded denim blue.

By sunset every gutter and ditch brimmed with rainwater hurrying seaward. The Cobequids wore their usual head of evening cumulus, but the continental air had swept the rain clouds toward Cape Breton and Newfoundland. The glass was rising. In the raking orange light, things looked decidedly fallish. By dusk the thermometer threatened frost. As darkness fell and a full moon floated above the eastern horizon, I went out and draped a bed sheet over the tomato transplants just in case.

And today, a tremendous lightning/thunder storm from 9:30 P.M. to 11:30 P.M. Sheet lightning lit the sky from east to west, chain and fork lightning too; so fast, yet frozen in time on the retina. Such a glorious racket! I turned off the power for fear our wires would melt. In the midst of it my brother called, but I, recalling how lightning had once set fire to our basement via the telephone line, told him to hang up and I'd call him back.

July 5

In an old anthology I bought at the University Women's Annual Book Fair I found this 1940s poem by Carl Sandburg in which he compares the beauty of people to the beauty of weather:

> The people is a polychrome
> a spectrum and a prism
> held in a moving monolith,
> a sensole organ of changing themes
> a clavilux of color poems
> wherein the sea offers fog
> and the fogs move off rain
> and the labrador sunset shortens
> to a nocturne of clear stars
> serene over the shot spray
> of northern lights.

And in Dorothy C. Fisher's *The Night on the Cobble*, I came across this: "The sun thrust one fiery shoulder over the mountain, and all the world gave a shout of colour." I like that. It reminds me of a sunrise I saw this week.

August 27

At last, some rain, a shower at 7:30 P.M. that lasted half an hour, falling straight down. The crabapple leaves twitched as if a small bird was moving among them. (At first that's what I thought it was.) No wind, just the plash of large individual drops on dry, firm leaves. A dance of gratitude: "The gentle rain that loves all leaves."

September 30

Again this year, noticed little flattish webs on the lawn, each sequinned by sunlight sparkling on dew. Averaging 30cm/1ft apart (though sometimes touching to form one oblong tent), each is about the size and shape of a rather deep tea saucer. The web is fastened to two or three blades of short grass which the spider has bent inward like fishing rods. Every second or third one has a small (uneaten) fly entangled low down. A couple had what looked like an egg, 1 mm long by 1/2 mm wide, yellow-gold with internal dots (eyes?). A parasitic wasp's egg? Never saw the webs' owners, even when I jiggled the webs. Probably still in their burrows, stiff from the overnight chill.

October 11

...last night, as I was finishing a display at work, a howling nor'easter sprang up, spitting cold rain. This morning the same gale gave snow for an hour. Afterwards the sky was magnificent: pillared clouds and pinnacles, dark on light, light on dark, all on windswept blue, with quick cold showers every few minutes.

Fall weather is so fickle and blustery, going from sun to cloud to rain and sun in hours. A lot of east wind, a lot of moisture. "After August, firewood don't dry" runs the old saying. Leaves falling fast now, pattering like rain on the studio roof. Strong W to NW winds, cold, for three days; a lot of low cloud. Blew the greenhouse plastic loose for the first time since last year.

November 19

Fall hurricane. Warm rain, warm wind gusting from SE and S. Full moon, high tide. Air very clear the day before—"When far hills loom, expect rain." Saw at 4 P.M. a sun hound, briefly. Before going to bed I tied a tarp over my wagon-load of dry split firewood. In the night it rained heavily again, a downpour from the warm front.

Next morning cool but sunny. In the night, first real ice. A centimetre thick in curbside puddles; on John's trout pond strong enough to skip pebbles across. High cirrus coming in from the south after a 4°C night, warm, moist ocean air sliding gently over cold, dry continental air until its moisture is wrung out and Gulf of Mexico air shoves the Arctic air aside. And then the reverse....

December 2

Falling behind in my journal. Ground hard with frost, fields pale green fading to tawny yellow. Our gravel road, frozen and too cold to melt snow, is dusted with it. Distant leafless birch and maples look purply-pink, but in the white light the dark spruce and fir look sombre and huddled from the cold. Shrubs bare, sky gray. Blue jays, having rifled the last sunflower heads in the garden, peck the few yellow apples left on our Bishop Pippin tree.

Overnight the wind veered from NE to N to W, becoming colder and colder until we're back to -10°C. Today dawned clear, with lovely bright light on the fluffy snow. Stove eating a lot of wood—hard to keep box full. Accompanied Jim and his sister Barbara and her husband Martin to the woodlot to find two Christmas trees. We did; they went away content.

After a stiff ESE to E wind with fine snow and then rain, we got two days of rushing Arctic air from NW and W. Blizzard in

PEI and northern New Brunswick, but we're to "expect flurries where winds cross open water"; only 5 cm here. Temp. minus 10°C and falling. The throttle on Fred's new half-ton 4WD froze and he had to leave it in our driveway while I took him and Julia and Shawna home in (ahem) our 1981 Chev Impala.

This morning, clear sky, slanting amber light on tree and patio, sky pale turquoise, bearing no resemblance to yesterday's scowling grey. Toward sunset, faint cobalt blue undertones. The sun sets noticeably later and rises earlier now.

Daughter Joyce and I skated on the meadow pond, and played hockey with a ball and two pucks. My stick was right-handed but I shoot left. Sidney our dog didn't care for the smooth ice, but Murph the cat came out to watch and even chased the puck a little.

22

Global

Warming

& Us

Everyone talks about the weather, but no one does anything about it.
—Charles Dudley Warner, 1897
(Often incorrectly attributed to Mark Twain)

Two guys in a Gary Larsen "Far Side" cartoon are fishing on a lake when one of them suddenly notices mushroom clouds looming over the horizon. "Hey," he crows, pointing, "you know what those mean?"

"No, what?"

"No size restrictions, and to hell with the limit!"

On the subject of global warming, we're all a bit like those fishermen. We make jokes: "Good, we won't need to go south any more!" "Good, it'll help my heating bills!" Or "I'm for it; no more winter clothing to buy!" Myself, I've never had such good luck with muskmelons. Muskmelons take a lot of heat to come to full sweetness and flavour, but recent summers have been perfect. The same goes for cukes and tomatoes.

On the other hand, I've never seen so many apples fall before their time—a bushel a week from our three oldest trees. The main reason apples drop prematurely is worms, in this case apple maggots (called "railroad worms" because they leave a brown trail in the flesh), and codling moth larvae. Both insects thrive in warm weather and I don't spray any more. I just pick up the drops and cart them to the woods for the rabbits and deer.

All trees have been under extra stress lately. In Truro, for example, elms that had been rescued from Dutch elm disease through the careful pruning of dead limbs succumbed to weeks of sweltering weather. A tree's way of staying cool is to step up transpiration—something like a dog's panting. The extra effort left the elms no reserve energy to fight the spore-laden beetles nor the attacking fungi. Shrubs that would have been lush and green turned brown midseason.

When, as in 1999, warm temperatures held into late fall, mayflowers mistook it for spring and bloomed in mid-November. If this trend continues, many cherished weather sayings will no longer hold true, and new ones will have to be invented.

Or we could dismiss the whole thing as a blip in the planet's long history of weather ups and downs. "If global warming is true," wrote a gardener friend of mine, "how come my tomato plants froze last May for the first time in forty years of gardening here?"

We've all heard about it, we know it could be serious; but it's too far-fetched, too mind-boggling, too, well, *global*, to take seriously. And yet....

Alarm Bells

In the 1970s few reputable scientists believed in global warming, even though alarm bells been rung years before by Nobel Prize-winning Swedish chemist Svante Arrhenius. Arrhenius, without a computer and using crude models, correctly predicted the current temperature of the earth. At the time, his was a voice in the wilderness. Today he'd be welcome at any meeting of serious climatologists. And if he asked for a show of hands as to who believed his theory, roughly ninety percent would support him.

World leaders, normally preoccupied with matters of state, are starting to pay attention. They are convening mammoth conferences where they promise each other that they will cut carbon dioxide (CO_2) emissions by such-and-such in so-and-so many years. The Kyoto Protocol of 1997 comes to mind, the failed November 2000 Climate Summit and the Bohn Summit of July 2001, where Canada re-endorsed its Kyoto goals.

Even ordinary politicians are mouthing the correct buzzwords and nailing weather planks to their election platforms. And in 1996 a consortium of American insurance companies, fearing weather-related disasters would "bankrupt the industry," sponsored a study of the possible role of greenhouse gases. German and Swiss companies have lobbied for reduced emissions. About the only ones who aren't convinced are the oil-producing nations and the petrochemical industry.

Something *is* up. But what is it?

The Greenhouse Effect

Literally, CO_2 is up. The air above us contains more CO_2 than it did in pre-industrial times. And although this gas can be poisonous—a fact which submarine crews never forget—that's not what the fuss is about. The fuss is about a side effect. It so happens that CO_2 traps heat—i.e., its molecular structure allows the sun's infra-red radiation (heat) in but won't let it out again. Methane (or swamp gas; CH_4), the second culprit in this drama, does the same. It too is increasing as garbage dumps, cattle feedlots, and manure ponds proliferate. These two invisible gases, along with nitrous oxide and chlorofluorocarbons (CFCs), act like a sheet of plastic suspended over the whole globe. The phrase "greenhouse gases" is apt.

The result seems to be a minuscule but steady warming of the earth's surface and everything on it. The air itself isn't warming any more than window glass heats up when sunlight passes through it. But the sunlight heats objects in the room.

But isn't CO_2 a natural part of the earth's atmosphere? It is, but as we saw in the chapter on clouds, the percentage is so tiny—only five parts per thousand, less than a speck of pepper on a large pizza—as to seem insignificant. In fact it's just the right amount. Circulating through the incredibly complex carbon cycle, it has been sufficient to help foster life on earth despite bouts of volcanism and glaciation, and to help maintain the apparently delicate global temperature regime.

What has upset the balance? The prime suspect is the accelerated burning of coal and oil and natural gas during the last 250 years. Ever since Thomas Newcomen and James Watt perfected the first heat engine in 1712 and 1769 respectively, and ever since Michael Faraday demonstrated the nature of electricity in 1826 and invented the dynamo, we've been on a fossil energy binge. Factories, trains, steamships, mining, motor cars, modern highways, aircraft, efficient home heating, plastics—all proliferated from those discoveries, and are still doing so.

Before the seventeenth century, most natural carbon was geologically locked away underground or undersea. Originally it had come from the dead bodies of plants (coal) and animals (oil and gas), from bodies both tiny (plankton) and large (e.g., giant ferns and horsetails). Alive, the plants in turn had taken the carbon from the air—incidentally reducing the primordial CO_2-rich

Canadians are the world's second largest emitters of greenhouse gases. The Americans come first, accounting for 18 tonnes of CO_2 per person per annum.

atmosphere to something breathable by humans. So long as those carbon deposits remained untouched, the balance held.

Once burned, however, carbon releases more than heat. As leftover carbon (C) from combustion and volcanoes has done for eons, it bonds chemically with existing atmospheric oxygen (O_2) to produce atmospheric carbon dioxide—in addition to what was already there. It's been doing that at an accelerating rate, first in Britain, then on the continent and in North America, for over two centuries. Lately the same thing has been escalating in the Third World, whose people naturally want to enjoy the benefits we in the west have come to take for granted.

To help pay for those benefits, tropical rainforests have been cleared at an unprecedented rate, using fire as the chief tool—more CO_2. Forest fires have always spewed out CO_2, but not on a global, ongoing, day-and-night scale such as we now see. Scientists are telling us that the earth's built-in CO_2 scrubbers, namely forests (especially tropical forests, as boreal forests are virtually dormant for half the year), marine phytoplankton, plus algae and aerobic bacteria, can no longer handle it all.

So we have this thickening film of insulating gases. The trapped heat becomes part of the planet's existing "body heat," generated not only by unaided sunlight but radiated from its molten iron core and from natural radioactivity. The resultant increase, scientists now claim, is measurable. For example, the late American astronomer-writer Carl Sagan said in his 1997 essay "Ambush: The Warming of the World" that overall, the earth's temperature barely increased in the twentieth century. He then went on to say that the 1980s and 1990s were the ten hottest years since 1860—despite the cooling effect of the 1991 explosion of the Philippine volcano Mount Pinatubo.

PINATUBO

Pinatubo's contribution of 20-30 megatonnes of sulfur dioxide and aerosols did cool things off worldwide for three years, but by 1995 the warming trend had reasserted itself. The countdown to 2000 was marked by record warmth worldwide. "We've seen fourteen of the warmest years on record occur in the past eighteen years," said David Phillips, a Toronto-based senior climatologist with Environment Canada.

If we discard the hype that surrounds all scary theories (remember Y2K?), we're still left with a core of evidence that's hard to wish away. An overwhelming majority of climatologists—American, British, German and others—now concur that if the trend continues, a worldwide warming of 1°C to 4°C will take place during this century. They also state that, even if we curbed CO_2 and methane emissions today, twenty years would pass before any decline could take hold, and decades more before the atmosphere returned to normal. Like an ocean liner, it will take a long time to stop.

A few degrees may not seem like much, but no one knows exactly what the effect will be. As Sagan pointed out, "The typical temperature difference for the whole world between an ice age and an interglacial interval is only 3 to 6 Celsius degrees (5 to 11 Farenheit degrees)." And no one knows for sure whether there is a link between global warming and ice ages.

CO_2 & Ice Ages?

One scientist who thought so was the late Patrick McTaggart-Cowan, former head of the Canadian Meteorological Service. In 1974 he warned that Canada should look to its vulnerable transportation network—we have more highways per capita than almost anywhere else in the world. And that we should start worrying about hanging onto our southern agricultural land, and stop paving it over.

No one knows for sure what triggered the ice ages even today. But no one disputes that they did happen, and that glaciers start with excess snow and cold. The fact is, enough snow fell to bury most of the northern hemisphere. When the snow got thick enough it turned to ice, which went on thickening until it began to flow like icing poured on a cake.

From Kansas north, the ice left its footprint on the land's face. Not only did it depress the land many metres—one hectare of 1.6km/1mi-thick glacier weighs over 17 million tonnes—it sculpted the surface in countless ways. It allowed the sea to invade river valleys, it bulldozed lakes out of existence and created new ones, it left boulders perched on mountaintops.

Hardly less powerful was the effect of meltwater floods that deepened old river channels and carved new ones on its way to the sea. Most of our soil ended up offshore as shallow "banks." But the land's loss was the sea's gain, for those rich sunlit waters nourish the plankton that ultimately feed all other forms of marine life.

Not many scientists expect another ice age, but now few doubt the link between global warming and rising CO_2 emissions. Millennia-old ice cores lifted in the 1990s from the Greenland and Antarctic ice caps prove that. And because CO_2 from natural causes leaves a different nuclear "signature" than CO_2 from human causes, it's possible to tell them apart.

Carbon dioxide from volcanic eruptions, wildfires, and decomposition of plants and animals has always been present in

GLACIAL BOULDERS

Somewhere between 2009 and 2012 global oil production[,] after rising steadily for 140 years, will peak and start down again. By 2020, about one-fifth of the predicted consumption will have to come from 'unidentified unconventional sources' (i.e., they have no idea where it's coming from.) By 2040 total global oil production may be down to less than half what it is now.
—Columnist Gwynne Dyer, Truro Daily News, Sept. 16, 2000

the atmosphere. Without its natural blanketing effect, the planet's surface would be 20-30°C/68-86°F colder and quite uninhabitable. Numerous intricate cycles have maintained the heat balance for over 600 million years. When the earth cools, CO_2 production is increased; when it warms, production is curtailed while CO_2 storage (in the ocean, for example) is stepped up. These cycles operate on a response time of millennia.

Most scientists are convinced that humans are part of the current rise. In 1995 the Intergovernmental Panel on Climatic Change concluded after exhaustive study by its twenty-five thousand scientists that "the balance of evidence suggests there is a discernible human influence on climate." Michael MacCracken, director of the U.S. Global Change Research Program, declared that the evidence "is becoming quite compelling." Thomas Karl of the U.S. National Climatic Data Center said the observed warming was "unlikely to be caused by natural variability," and that "there's a 90 to 95 percent chance that we're not being fooled." This is as close to an endorsement as we can expect from a scientist.

In March 1999, scientists at the universities of Arizona and Massachusetts announced the results of the reconstruction of earth's average temperature over the centuries. The shocking result was that a nine-hundred-year cooling trend has been suddenly and decisively reversed. Barring some error that everyone has overlooked, the odds are good the patient has a fever.

Still, it all seems a bit remote: invisible gases, remote ice core samples, faint chemical traces, complex computer simulations. What about harder evidence, something a non-scientist can see and touch, something that overrides the contradictions, the puzzling frozen tomato plants?

Harder Evidence

In the fall of 1999, CBC TV'S "Magazine" ran a special on the polar bears of northern Manitoba. It seems they are starving. This worries the people of Churchill and other coastal communities, people who have had more than their share of encounters with bears. Researchers found that these bears spend the short Arctic summer on land, where they give birth and raise their cubs. Normally they arrive fat from eating seals they have killed on the pack ice before breakup. This fat is supposed to keep them going despite meagre rations until the ice returns.

In recent years they have come ashore earlier than usual, looking thin and hungry. They are starving because Hudson Bay is

We're plugging in an electric blanket but we're not cold. What's worse, we're not taking effective steps to reduce CO_2 production it's like losing the thermostat after setting it on high.
—Dr Mike Apps, Northern Forestry Centre, Edmonton, Alberta, 1988

Despite the contributions of New Brunswick's Point LePreau atomic energy plant and Cape Breton's Wreck Cove hydroelectric project, most Maritime electricity still comes from burning coal and oil.

Canada's Mackenzie River Basin has been warming at three times the global rate, leading to melting of the permafrost and to extensive slumping of the river's gravel banks.

thawing earlier and freezing later. The earlier spring not only disperses the seals they feed on, but it forces the bears ashore earlier. As a rule, each increase of one degree Celsius moves spring one week ahead. When the bears come ashore, they find the ground strangely dry, even dusty. Normally it is boggy on top and solid ice a metre down.

And with freeze-up coming later than usual, the huge bears must subsist that much longer on lemmings, grubs, and berries. By the time the nursing females get back to eating seals, they are malnourished and the cubs suffer. No wonder polar bears have been raiding the Churchill town dump in recent years.

In the summer of 2001, several Canadian cities noted an upsurge of rats due to mild winters and hot summers.

In July of 2000, for one of the few times in living memory, the North Pole area became open water. Six years earlier, the area had floating ice 2-3m/6.5-10ft thick. A very rare event.

There was another warm spell a thousand years ago. Although the mean temperature increase was only about 1 Celsius degree, it was enough to enable Norse settlements to take root on Greenland's southwest coast. Reduced sea ice was likely what made such long northern voyages feasible, including their exploration of our coasts and their abortive attempt to settle here. The Greenland settlements were self-sufficient, and lasted four hundred years before permafrost resumed its lethal grip.

Even so, Norwegian researchers were not surprised that the ocean at the North Pole had thawed. As early as 1978, they were reporting a shrinkage in Arctic ice cover. This was confirmed in 1998 by an international group of scientists living for months aboard the Canadian icebreaker *Des Groseilliers*. In December 1997 the vessel was allowed to become frozen in the Arctic ice pack to study the pack's wanderings. The following summer they were shocked to find the sea ice less than half its normal thickness and melting fast.

At the opposite pole, enormous rifts have opened in Antarctica's Wordie Ice Sheet. And in January 1995, Antarctica's Larsen Ice Shelf calved an ice pan the size of Prince Edward Island. Four years later, an even larger chunk slid off the Rosse Ice Shelf in the southern Weddell Sea. Mountain glaciers around the world have also been retreating faster than normal.

Extreme Weather

Most disturbing of all, extreme weather is becoming more common. Bad weather can certainly happen without global warming, and global warming does not account for *all* bad weather. But it's a fact that the more extreme the air temperatures when cool and warm air currents collide, the more violent the resulting hurricanes and tornados. All computer models of future climate predict significant increases in bad weather, both hot and cold. They also point to severe drought inland, and severe flooding in coastal areas.

The years from 1996 to 1999 exhibited the most violent weather since records were first kept over a century ago. In Central America, Southeast Asia, and Africa, torrential rains, deadly mudflows, and crippling floods have dominated the news. Europe and North America have been hit by sudden and unseasonable snowstorms. Since 1988, the American Midwest has experienced droughts like those of the Dust Bowl years. According to some projections, the Midwest and the Ukraine—the world's two breadbaskets—will become scrub deserts if current warming trends continue. Not surprisingly, forest fires have increased in size and frequency across the northern coniferous belt. Montana's millennial summer was the worst for wildfires since records have been kept.

In 1998, Canada-wide statistics put forest fire incidence at 123 percent above normal, with the area burned 159 percent higher.

Even sub-tropical deciduous areas like Australia and Florida are being hit hard. My friend Jenny Mills and her husband, who live in Darlington, Western Australia, witnessed a big brush fire nearby on New Year's Eve: "It burnt about one hundred acres—water bombers and helicopters flying overhead!" In February 2001, Florida suffered the biggest forest fires in a century.

Weather-related disasters are costly. In 1998, the Worldwatch Institute in Washington D.C. declared that the first eleven months of that year set a record for weather-related disasters around the globe: $136 million in damage and economic losses, thirty-two thousand people killed, and 300 million people forced to resettle or leave their homes. Among these disasters was Hurricane Mitch, which killed eleven thousand people in Central America in October; China's devastating summer floods along the Yangtze River; and India's heat wave in which twenty-five hundred perished.

Record flooding in southern Vietnam transformed rice fields and rural communities into "desolate lakes," the Red Cross said Friday.... Heavy rains began in July, a month and a half ahead of the normal monsoon.

—*Truro Daily News*, Sept. 16, 2000

Canada has also had its share of strange weather. In 1996 Quebec's Saguenay River burst its banks, leaving two thousand families homeless. December 1997 ushered in a twelve-month period of abnormally warm weather, thought to have been caused by warm water upwelling in the Pacific—the famous El Niño effect. As a result, summer water levels in the Great Lakes and in the upper St. Lawrence River fell twice as far below their spring peak as usual. Container ships approaching Montreal had to lighten their loads to avoid touching bottom. In the same year, Winnipeg's Great Flood drove twenty-eight thousand people from their homes. The Red and Assiniboine rivers have

always flooded every few decades, but this seemed worse.

In January of that year, southern Ontario got a winter's worth of snow in two weeks. Much to the amusement of snow-wise Maritimers, Toronto panicked and called in the army to help clear the streets. On Christmas Eve, 1998, Vancouverites slogged through their second rare blizzard. The winter of 2000-2001 gave Manitoba and Saskatchewan their deepest snowpack in living memory.

THE GREAT ICE STORM OF 1998

Then there was the infamous Great Ice Storm of January 1998. As winter storms go, it was well behaved. There were no gales and no blizzards, just several days and nights of mist and freezing drizzle. Yet the cumulative ice buildup crumpled hundreds of hydro towers across western Ontario, eastern Quebec, and the East Coast, plunging three million people into the cold and dark for weeks and killing twenty-five. It also mangled tens of thousands of trees.

Possible Scenarios

What is the worst that could happen? If the scientists are wrong, the earth will simply revert to normal and we will have alarmed ourselves for nothing while amassing some fascinating data. If they're right, the planet's inhabitants can expect steadily worsening weather and related disasters; an increasing incidence of "tropical" diseases; major losses of habitat, cropland, and wildlife; more heat waves and cold snaps, more avalanches in mountainous terrain, and more coastal erosion; more shrunken water tables and salty drinking water, even.

As the average temperature continues to rise, sea level can be expected to rise too. That's because huge amounts of water are locked in polar ice and mountain glaciers. Ice isn't like snow, which is so fluffy you have to melt ten units to get one unit of water. Ice has virtually the same density as water. That's why an ice cube floating in lemonade barely breaks the surface. A melting iceberg with a top the size of a cathedral adds that much

Eighteenth-century boat mooring rings at Fortress Louisbourg in Cape Breton are now a metre under water at high tide.

water to the ocean.

During the last century, sea level increased by approximately 1.5cm/0.6in per decade. Some of this may have been due to the natural sinking of coastal regions as northern land masses rebounded from carrying loads of glacial ice up to 5km/3mi thick for scores of millennia.

But that is a slower process, and geologists think such *isostatic* sinking has likely ceased this far south. Most of the recent rise seems due to warmer weather. Oceanographers calculate that two-thirds of the rise is due to thermal expansion, while the other third is from meltwater released by glaciers and icecaps. Since all the world's saltwater seas are connected, ice melting anywhere eventually raises the whole global ocean.

The expected increase during this century is 20cm/8in to 1m/3.3ft. A one metre increase may not seem like much. After all, the natural daily tidal range is about two metres. Worldwide, however, the disruption and economic loss would be colossal. Such an elevation of normal high tide would inundate the lower deltas of most of the planet's major rivers such as the Nile, Ganges, Mekong, Amazon, and Mississippi. New Orleans, awash twice a day, would look like Venice. (Venice is used to it.) During storms it would be worse. A deeper ocean would also flood the downtown streets of coastal cities such as Boston and Copenhagen. San Francisco and Vancouver would largely escape, being hilly. So would Paris, being well upriver; but east

- - - - EXISTING
SHORELINE

NEW SHORE
LINE!

HOW THE MARITIMES MIGHT LOOK
IF THE SEA ROSE 100 m /328 ft

Londoners on the tidal Thames, and downriver cities like Greenwich and Gravesend, would get their feet wet every high tide.

Our seaside cities—Saint John, St. John's, Halifax—might sustain less damage because most were built on hillsides, though low-lying places like Yarmouth and Charlottetown would certainly be flooded. Not so for Fredericton, which lies far upriver. Moncton is a river town too, but Fundy tides already flood the lower Petitcodiac twice a day. It is doubtful whether its earthen dykes would restrain even higher tides, especially at the new or full moon with a strong westerly wind and low atmospheric pressure. The same applies to most Bay of Fundy centres, especially those at the inner extremities, such as Sackville, Amherst, and Truro.

For that matter, such dykes protect thousands of hectares of Maritime farmland reclaimed from the sea during and since Acadian times. Whether they could withstand even higher tides, perhaps driven by gale force winds and ice, is unknown. Certainly erosion of beaches and estuaries, already a fact in some areas, would increase dramatically. Port authorities caught unprepared would have to spend a lot of money to counter the threat. Even if global warming were halted now, say experts, the global ocean would keep rising *for a century or two*. Canadian weathercaster David Phillips has jokingly suggested that investors should buy into companies building Maritime sea walls.

Armchair scientists caused a stir in the 1980s by speculating that recent ice ages may have been brought on partly by the Gulf Stream spilling into the Arctic Ocean, causing massive snowfalls farther south. Rising sea levels, they said, could allow that river of warm water to breach a supposed shallow barrier between Iceland and Scotland.

Other suggested causes of ice ages include precession of the equinoxes (the earth's wobble), a drop in the sun's output, and sunspots.

Global Superstorms

The scariest scenario is a global superstorm triggered by a sudden shift in ocean currents or jet streams or both. Erratic jet streams are also powerful engines of weather change, as victims of sudden Florida blizzards know only too well.

In *The Coming Global Superstorm*, Art Bell and Whitley Strieber paint a terrifying picture of such a scenario, and they blame global warming.

Here and there throughout the northeast there are raised beaches (marine terraces) testifying to a time when waves broke on beaches as much as 100m/328ft higher vertically than they do today. There are also drowned forests, with stumps still rooted where sea mud buried them millennia ago.

In 2000, a Gulf of Mexico squid turned up in Newfoundland's south coast waters, suggesting that the Gulf Stream had swung inshore. At the same time, several rare southern bird species showed up at St. John's feeders.

The book focuses on our hemisphere because that is where the last few glaciations started, and because the southern hemisphere is mostly ocean. They expect the superstorm to come sooner than later. They predict a proliferation of unusual isolated weather systems that in time coalesce into a few enormously powerful systems. These, they say, will drag Arctic air into southern latitudes, creating killer cold waves that instantly freeze everything in their path. They see mass migrations of humans southward.

If this happens in fall or winter, they add, the massive sudden buildup of snow and ice, by chilling the air and reflecting solar heat back into space, could linger into the next summer and the next, bringing on a new ice age. They also claim that such storms have happened in the past more than once, and suddenly. And that in one case it happened in early summer, leading to enormous floods of meltwater that crammed now-frozen Arctic ravines with the jumbled remains of frozen animals, many of them southern species, and many of them with undigested food in their stomachs and even in their mouths. In one case, they say, diggers found a budding apple tree that had been entombed in permafrost for thousands of years. How else account for such anomalies? Immanuel Velikovsky posed the same question in the 1950s.

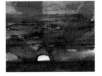

Well, it's one way to sell books. Yet much of what they say is true. Even if they're only half right, there seems to be cause for alarm. A sudden shift to Arctic weather may not be as far-fetched as some think. In late August 1996, the internationally funded, Halifax-built research vessel *JOIDES Resolution* was drilling seabed cores in uniquely well-preserved layered sediments in the 230m/754ft-deep Saanich Inlet on the British Columbia coast. The crew chose this site because shallow water at each end prevents normal tidal flushing. Since few aerobic organisms can survive in its oxygen-poor waters, bottom sediments are virtually undisturbed.

Richard Hebda, a Canadian researcher who examined the first cores, witnessed a startling discovery. One of the first 9.5m/31ft sausages of mud clearly showed such a shift. About eleven thousand years ago, average temperatures on the West Coast plummeted from near modern-day values to North Pole temperatures in just a few years. "Geologically speaking," said Hebda, "it is an instantaneous change of the climate."

Though these findings are deeply disturbing, no one is certain what they portend. Global superstorm predictions notwithstanding, we're almost sure to see a continued warming, espe-

Incoming solar energy must equal heat loss from the earth, or the system would overheat or freeze. Luckily it is almost exactly balanced, with 51% absorbed by land and sea, 19% by the atmosphere, and the rest radiated back. Without the presence of atmospheric moisture and ozone, equilibrium would still prevail—but at an average global temperature of -20°C/-4°F.

cially in polar and temperate latitudes where the atmosphere is thinner. The melting of northern ice will reduce the planet's reflective surface or *albedo*, thereby speeding heat absorption.

Planetary heat absorption varies with surface cover. Forests trap as much as 97 percent of incoming radiation, water 90 percent, grassy fields 70-90 percent, sand 55-85 percent, thin clouds 50-70 percent, fresh snow 5-25 percent, thick clouds as little as 5 percent.

A global warming scenario will see vegetation edge northward: grassland will invade taiga (the northern coniferous zone), taiga will move into tundra (treed bog and barren), tundra will overtake muskeg, muskeg will colonize areas of dwindling permafrost and former icefields.

A warmer climate will be a drier climate. This will kill vegetation unable to migrate fast enough. With nothing left to anchor the formerly frozen soil, drying winds may raise dust storms of silt and peat that will blanket large areas in loess, as they did in China and elsewhere after the last ice age.

In the nineteenth century one of the noisiest debates, apart from Darwinian evolution, was the question of ice ages. One side maintained there was no such thing, that the so-called proofs—massive boulders of foreign composition perched on hilltops, scored and polished granite pavements, flocks of whaleback gravel hills all facing the same way—were caused by a universal flood. European glaciologist Louis Agassiz patiently proved them wrong—but not entirely. Glaciers, since they do depress the land, cause massive flooding.

Likewise our fears about global warming merit being taken seriously. For this crisis, if real, ranks with nuclear war in its potential to harm life on earth.

Fortunately, the scenario is reversible. The gravest folly would be to "wait and see." To use a barnyard analogy, it's no use shutting the barn door after the horse is gone.

A Very Special Oxygen

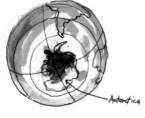

OZONE "HOLE" OVER THE SOUTH POLE

No discussion of air, even a brief one, would be complete without a note on the importance of ozone (O_3), a form of oxygen with three atoms instead of two, a sort of overweight oxygen. It is a blue gas and is the source of the sulfurous odour one smells when model trains or welders make electrical sparks, and sometimes after electrical storms. Down here where we live it's a pollutant, but at high altitudes, say 20-25km/12-15mi, it forms a vital shield protecting all life against deadly ultra-violet and other solar radiation. Along with water vapour, it also helps maintain the delicate balance between heat gain by surface absorption and heat loss by radiation.

Normally, the production and loss of natural ozone also stays in balance. Human activity has been jiggling that balance for decades. One cause is ozone-destroying chlorine-based chemicals called chlorofluorocarbons (CFCs). CFCs are found in older refrigerants, in solvents for cleaning electronic circuit boards, in aerosol propellants, in soft foams used in packaging and insulation, and in halons used in some fire extinguishers. Sulfur dioxide from volcanoes, pulp and paper mills, and vehicles also destroys ozone.

In 1986 researchers who went to Antarctica with the U.S.-sponsored National Ozone Expedition named chlorine and bromine as the main culprits in the destruction of up to 40 percent of the South Pole's stratospheric ozone shield. After a series of high-altitude flights in 1987 confirmed this, a worried United Nations adopted the "Montreal Protocol on Substances that Deplete the Ozone Layer" that September. The worry was that ozone-poor air would be carried worldwide, diluting normal air.

[Even] with the best efforts, CFCs and ozone-destroyers already in the air will continue eating away at the ozone shield well into the next century.
—USA Today's *The Weather Book*

The agreement called for a 50 percent reduction in CFC production by the year 2000. In 1990 scientists discovered that the Arctic was actually losing its ozone shield. Alarmed world leaders agreed in London to speed up reductions in CFC emissions and to ban all CFC production—but not use—by the year 2000.

In 1997 Canada signed the international Kyoto Protocol, promising to reduce greenhouse gas emissions by six percent below 1990 levels between 2008 and 2012. The new millennium dawned, finding the provinces polluting the atmosphere at about the same rate, though the federal government had pushed hard for change. Predictably, the worst polluters are the most industrialized: British Columbia, Alberta, Saskatchewan, Ontario and Quebec. Together they puff out ninety percent of the bad gases. Atlantic Canada contributes barely ten percent, but per capita we're almost as bad. In October 2000, the first ministers met in Quebec City to seek agreement on a national strategy. But at the Summit of the Americas in May 2001, they and other western leaders were chagrined to witness new U.S. President George W. Bush pull his country, the world's number one air polluter, out of the Kyoto Accord.

If UV radiation continues to increase, the incidence of skin cancers, especially facial cancers, will escalate. While sun-blocking creams do help prevent sunburn, few people will wear them year-round. One would expect that thin-skinned creatures like amphibians and young fish, which hatch and live in shallow

Every one-percent decrease in ozone concentration could increase ultraviolet radiation by two percent. This could lead to nearly fifty thousand new cases of skin cancer throughout Canada every year, including deadly melanoma.

sunlit water, are especially vulnerable. In fact, since the late 1990s scientists have noted major declines in frog and toad populations worldwide. National frogwatch programs have been set up to find out why. If the frogs are at risk, so are we.

A more insidious threat is the possibility of killing off the microscopic surface-dwelling plankton that swarm in the world's oceans. Plankton are not only major consumers of carbon dioxide, but they fuel the oceanic food web, which in turn provides much of our food. When a codfish eats a herring, it is indebted to the plankton that fed the capelin that fed the herring. The same applies to everything from sea slugs to stormy petrels to sperm whales.

Fortunately, the reduction in CFC emissions already seems to be having an effect. There are signs of a slow-down in ozone depletion. People can speed the recovery by refusing to buy or use any product not yet free of CFCs, and by pressing for CFC-free substitutes. We need that ozone shield.

In a late episode of *Star Trek*, Captain Picard is shown wearing a wide-brimmed hat and preparing to depart their UV-baked planet. Could it happen on planet earth?

What to Do?

There are some things we can do to reduce or reverse global warming and ozone depletion.

Get Informed Apart from the Weather Network (www.theweathernetwork.com) and the Weather Channel (www.weather.com), the Internet is full of useful (and not-so-useful) weather information. Among the best are Environment Canada's "Green Lane" (www.doe.ca/weather_e.html) and the CBC's web site (www.cbc.ca), as well as the David Suzuki Foundation's web site (www.energyrevolution.net). Public librarians can also direct you to good books and articles.

Along the way, try to convert your family and friends, especially the young. But don't just talk, do! Make some lifestyle changes. Here are a few green ideas:

• Go back to using one motor vehicle (or none at all); drive slower.

• Don't buy or drive gas-guzzling SUVs, which escape the more stringent automobile emission codes by being classed as trucks.

• Use mass transit wherever possible.

• Vacation by ship or train, not by car, and especially not by travel trailer.

• Copy the Brits; wear more clothes indoors. If you have electric heat, install more thermostats, maintaining full daytime heat (20°C/66-68°F) only in the room(s) most used. At night turn all of them down about 5°C/10°F. Install ceramic heat storage units.

• Install or have your landlord install some solar panels on the roof.

• At stoplights, don't idle a car motor longer than 45 seconds.

• Cut down on the use of air conditioners.

• Turn off appliances, lights, and other electrical devices that no one is using.

• Upgrade insulation of attics, basements, doors, and windows to modern standards.

• Mow and water the lawn less often, and shrink it by planting shrubs, wildflowers, or a garden. By setting the mower higher (say 8-10cm/3-4in), it will stay greener and withstand dry spells better.

• Grow and store more of your own food so it won't have to be trucked from Florida or California or shipped from Israel or Africa; even a window box of lettuce and radish will help.

• Plant trees (especially urban trees), or support people who do. Trees cool hot pavement by evaporation and by providing shade.

• Boycott energy-gobbling super-malls with their area lighting and huge floodlit parking lots which remove plant cover, lead to lowland flooding, and encourage motorized shopping. For long-distance shopping use e-commerce and mail order.

• Reduce paper and plastic consumption by not buying over-packaged products, by reducing household subscriptions, and by refusing junk mail. Reuse, recycle, reduce.

• Buy in bulk whenever possible, and bring your own cloth shopping bags or basket.

One 100-watt incandescent light bulb left burning overnight may not seem like much, since it costs only a few extra cents a day and burns perhaps only a teaspoon of oil. But 100,000 bulbs and fluorescent fixtures left burning unnecessarily can greatly increase oil consumption, air pollution, and CO_2 buildup.

In January 2001, California's Silicon Valley, nerve centre of North America's computer industry, lost millions of dollars in a state-wide series of "rolling blackouts" designed to conserve increasingly expensive electricity supplied by private companies, some of whom buy from Canada.

- Reduce your use of aluminum (e.g., pop cans) and other metals. Bauxite, from which aluminum is extracted, takes huge amounts of energy to process. Recycle what you do use.

- Eat less beef, the raising of which requires clearing large areas of forest (which absorb and store carbon dioxide), and produces methane, a bad greenhouse gas.

- Put up a clothesline; fresh air is free, sunlight is germicidal.

- Unless you're running an institution, use a cold-water detergent and don't wash linen after each use.

- Move closer to your work, or work from home.

- Take shorter showers and/or install a water-saving shower head.

- Insulate your water heater and turn its thermostat down below scalding hot—say, 60°C/140°F.

- Buy local produce and crafts whenever possible.

- Cook in bulk and freeze the extras.

- Finally, get politically involved. Lobby your MPs; urge Ottawa to live up to its Kyoto promises. For climate change action campaign material, call 1-800-453-1533.

What if this whole global warming thing is a figment of overheated imaginations? We still won't have wasted our time, for a healthy environment will never be irrelevant.

Acknowledgements

As a landscape painter and nature writer, I've been jotting weather notes for years. But I had never thought of making a book on weather until Dorothy Blythe of Nimbus Publishing suggested it. She did so on the strength of a little essay I wrote about a thunderstorm. At the time, I was busy finishing a highway nature guide for Nimbus and couldn't tackle it. But I could see the merit of her suggestion. After all, who talks more about the weather than Maritimers?

In a sense, my skywatching career began in childhood. I owe a debt to my riverman father's stories, which were full of wind and ice and rain and snow. Later the Danish forester and artist Hans Mandøe mentored me in art and natural history. I owe another sort of debt to my longtime neighbours Carman Forbes, Bob Guild, and Ted LeMaistre, each a skywatcher in his own way.

Favourite writers have helped me too. I can never read the Welsh poet Dylan Thomas without thrilling to his weather. The same goes for Gerard Manley Hopkins and Robert Frost, and for the prose of New England's E.B. White. White's suspenseful "The Eye of Edna" is one of the finest weather pieces I know. Novelist Joseph Conrad's *Typhoon* is the ancestor of every modern weather epic.

Among those who helped more directly, a special thank-you to Bermuda meteorologist David Forbes (son of Carman), who directed me to likely local sources, including the works of Rube Hornstein, the popular Halifax-based weatherman of the 1970s and 80s. And to Peter Coade, ATV weathercaster.

Fellow forester David G. Dwyer furnished a copy of his monograph on the history of forest blowdown in Nova Scotia. My hiking partner Dave Taylor was a good sounding board as the writing progressed. Near the end of the project, Bob Mingo lent me some of his rare Eric Sloane weather books. I already owned three of this writer-artist's books on pioneer tools and lifeways, but hadn't known that Sloane was North America's first television weatherman—a contemporary of the CBC's colourful Percy Salzman.

All these people helped. So did the crusty farmer who chastised me for daring to forecast the weather over *his* fields. He didn't know it, but he lit a long fuse. Dammit, the weather belongs to all of us!

Resources

Not yet being nimble on the Net, I mostly made do with old-fashioned print materials such as my trusty dictionary and encyclopedia, many magazine and newspaper articles, and these books:

Bartlett, John. *Familiar Quotations*. Ed. Justin Kaplan. Boston: Little, Brown & Company, 1992.

Bell, Art, and Whitley Strieber. *The Coming Global Superstorm*. New York: Simon & Schuster, 2000.

Brown, Craig, ed. *The Illustrated History of Canada*. Toronto: Lester Publishing Ltd., 1996.

Bruce, Charles. *The Channel Shore*. Toronto: MacMillan Canada, 1954.

Burns, Robert. *Poems and Songs of Robert Burns*. Ed. James Barke, London: Collins, 1955.

Canadian Global Almanac. Toronto: MacMillan Canada, 2000.

The Climate Chronicle. Vancouver: David Suzuki Foundation, 2001 (summer issue).

The Climates of Canada. Ottawa: Environment Canada, Canadian Government Publishing Centre, 1990.

Conrad, Joseph. *The Mirror of the Sea: Memories and Impressions*. London: Methuen & Co Ltd., 1935.

Cook, Greg. *Love From Backfields*. St. John's: Breakwater, 1980.

DeBlieu, Jan. *Wind: How the Flow of Air Has Shaped Life, Myth, and the Land*. New York: Houghton Mifflin, 1998.

Duley, Margaret. *The Eyes of the Gull*. Toronto: Griffin Press Limited, 1976.

Forsdyke, A.G. Weather and Weather Forecasting. New York: Bantam Books, 1971.

Geddes, Gary & Phyllis Bruce, eds. *15 Canadian Poets Plus 5*. Toronto: Oxford University Press, 1978.

Gibbs, Robert. *A Kind of Wakefulness*. Fredericton: Fiddlehead Poetry Books, 1973.

Hopkins, Gerard M. *Poems and Prose*. Ed. W. H. Gardener. London: Penguin Books, 1985.

Johnston, Wayne. *The Story of Bobby O'Malley*. Ottawa: Oberon Press, 1985.

Kerslake, Susan. *Middlewatch*. Ottawa: Oberon Press, 1976.

MacLennan, Hugh. *Barometer Rising*. New York: Duell, Sloane, Pearce, 1941.

MacLeod, Alistair. *No Great Mischief*. Toronto: McLelland & Stewart, 1999.

MacLeod, Alistair. *The Lost Salt Gift of Blood*, Toronto: McLelland & Stewart, 1976.

Maillet, Antonine. *Pelagie*. Tr. Philip Stratford. Toronto: Stoddart, 1994.

Marine Weather Guide 1992; Environment Canada, Atlantic Region; *especially Scotia/Fundy, East Coast Weather Manual, Gulf of St. Lawrence*.

Marshall, Ingeborg. *A History and Ethnography of the Beothuk*. Montreal/Kingston: McGill-Queens, 1996.

Milne, A.A. *Now We Are Six*. New York: E.P. Dutton & Co., 1926.

Montgomery, L.M. *Anne of Green Gables*. London: L.C. Page & Co., 1908.

Mumford, Lewis. *Sticks & Stones: A Study of American Architecture and Civilization*. New York: Dover Publications, 1955.

Nelson, John L. *Practical Guide to Aviation Weather: Weather and Your Flight Plan Decision*. New York: Sports Car Press, 1976.

The Oxford Book of English Verse: 1250-1918. Ed. Sir Arthur Quiller-Couch. Toronto: Oxford University Press, 1939.

Phillips, David. *The Day Niagara Falls Ran Dry! Canadian Weather Facts and Trivia*. Toronto: Key Porter Books & Canadian Geographic Society, 1993.

Phillips, David. *Blame It on the Weather*. Toronto: Key Porter Books, 1998.

Pratt, E.J. *The Collected Poems of E.J. Pratt*. Toronto: Macmillan Canada, 1958.

Raddall, Thomas. *Hangman's Beach*. Garden City, NY: Doubleday, 1966.

Richards, David Adams. *The Coming of Winter*. Toronto: McLelland & Stewart Inc., 1992.

Sagan, Carl. *Billions & Billions: Thoughts on Life and Death on the Brink of the Millennium*. New York: Random House, 1997.

Sanger, Peter. *The America Reel*. Porters Lake: Pottersfield Press, 1983.

Thomas, Dylan. *Collected Poems*. New York: New Directions Publishing, 1971.

Thurston, Harry. *Clouds Flying Before the Eye*. Fredericton: Fiddlehead/Goose Lane, 1985.

Trefil, James, *Meditations at Sunset: A Scientist Looks at the Sky*. New York: Scribner's, 1987.

Weather. Loughborough, England: Ladybird Science Books, 1985.

Weather. New York: Life Science Library, Time-Life Books, 1970.

Operating just one of the new tar sands plants would produce the same amount of greenhouse gases as putting a million more cars on Canadian roads.

—Dr. David Suzuki

Weather Facts. London: Firefly Books, Dorling Kindersly, 1995.

Williams, Jack. *USA Today: The Weather Book*. New York: Vintage Books/Random House, 1992.

Hard to find, but useful and delightful for the weatherphile, are these books:

Hornstein, Reuben A. *It's in the Wind*. Undated. One of Canada's first weather booklets, issued by the Meteorological Division over 30 years ago; lovely woodcut illustrations, a collector's item.

—————. *The Forecast Your Own Weather Book*. Toronto/Ottawa: McLelland & Stewart/Environment Canada, 1980. This is more comprehensive, with lovely illustrations.

Kramer, Stephen. *Theodoric's Rainbow*. New York: Scientific American For Young Readers, 1995. How fourteenth-century monk Theodoric of Freiberg tried to explain the physics of rainbows two centuries before Newton; for ages K-4.

Minnaert, M. *The Nature of Light and Colour in the Open Air*. New York: Dover Publications, Inc., 1954. (reprint)

Sloane, Eric. *Almanac and Weather Forecaster*. New York: Duell, Sloan & Pearce, 1955. America's first TV weathercaster guides us through a year of weather month by month, with copious illustrations in his inimitable pen-and-ink style.

—————. *Folklore of American Weather*. New York: Hawthorn Books, (undated). An almanac of weather wisdom.

—————. *Look at the Sky...and Tell the Weather*. New York: Funk & Wagnalls, 1970. Sloane's favourite, it shows us weather through the eyes of a bush pilot, an old-time travelling parson, a parachutist, a farmer, a sailor, and so on.

While every reasonable effort has been made to obtain copyright permission for longer quotations (short ones have been credited in the text and their sources cited here), some may have been missed. The author would be glad to have such omissions brought to his attention for inclusion in a later edition.

Weatherspeak:
A Glossary

Absolute humidity The water vapour in a given volume of air

Air mass A large body of air with fairly uniform barometric pressure, temperature, and relative humidity

Altimeter A special type of aneroid barometer used in airplanes to measure altitude

Atmosphere The envelope of air surrounding the earth; most weather occurs in the bottom 10km/6mi

Atmospheric pressure The force exerted on the earth by the weight of air above a given area

Barograph A device for recording air pressure over time

Beaufort Wind Scale Used to classify wind speed based on visible events, devised in 1805 by British Admiral Francis Beaufort to describe winds at sea

Blizzard A winter event in which the wind speed is greater than 40kmh/24mph, visibility is 1km/0.6mi or less due to blowing snow for four hours or more. Temperature would normally be equal or colder than -3°C/27°F. Note: It does not have to be snowing.

Climate Long-term average that describes the kind of weather to be expected in a given region

Cold front A warm/cold air boundary with the cold air advancing

Condensation The change of vapour to liquid

Coriolis Effect The apparent curving motion of anything, such as wind, caused by the earth's rotation from west to east

Cyclone An area of atmospheric low pressure, from small dust devils to huge hurricanes, with winds circling around it, counterclockwise in the Northern hemisphere, clockwise in the Southern

Dew Point Measure of humidity expressed as the temperature at which dew begins to form

Downburst Wind blasting down from a thunderstorm or shower

Drizzle Falling water drops with a diameter of less than 5mm/0.02in

El Niño (Spanish for "Christ Child") Periodic replacement, near Christmas, of cold northbound Peru current with warm south-bound current, causing warmer winters in Western Hemisphere. The opposite is locally called La Niña.

Flash Floods Sudden rise of water with little or no warning, com-monly due to massive rainfall and/or snowmelt over a small area in a short time

Fog A cloud with its base on the ground

Freezing rain Supercooled raindrops that turn to ice when they touch something at or colder than the freezing point (0°C/32°F)

Front Boundary between air masses of different densities and usu-ally different temperatures

Frost Water vapour that has turned to ice on an object

Funnel Cloud Similar to Tornado but not reaching the ground

Gale Strong wind of 65-100kmh/39-60mph

Glaze A film of ice formed on an object when supercooled water spreads out on contact

Greenhouse effect Warming of a planet's surface caused by the absorption and re-emission of infrared energy by molecules in the atmosphere

Gulf Stream A warm ocean current that flows from the Gulf of Mexico northeasterly across the Atlantic to the coast of Europe

Hail Chunks of ice that form in thunderstorm updrafts

Halo A ring or arc of light around the sun or moon caused by ice crystals

Heat Wave A period with three or more consecutive days with temperatures at or above 37°C/90°F

High An area of high atmospheric pressure

Hurricane A tropical cyclone with winds of 120kmh/75mph or more

Hydrosphere The earth's water

Ice pellets Falling drops of frozen water 5 mm or less in diameter (Called *sleet* in the U.S., while in Britain the word means freezing rain mixed with wet snow)

Inversion Stable air condition in which air near the ground is cooler than air higher up (the reverse of normal)

Jet stream A narrow band of high-altitude wind with speeds faster than 91kmh/57mph

Land breeze Light winds blowing offshore when air over water is warmer than air over land; opposite of *sea breeze*

Lightning A visible discharge of electricity produced by a thunderstorm

Low An area of low atmospheric pressure

Maritime air mass A humid air mass, warm or cold, that forms over the sea

Microburst A downburst of wind less than 4km/2.5mi in diameter

Ozone hole The unexplained disappearance of ozone in the stratosphere

Probability of precipitation (POP) A subjective numerical assessment of the *chance* (not the amount, location, or duration) of measurable rain, drizzle, ice pellets, or snow at some time during the forecast period

Rain Falling water drops with a diameter greater than 5mm/0.02in

Rainbow Arc or circle of coloured light caused by refraction and reflection of light by water droplets

Relative humidity The ratio of the amount of water vapour actually in the air compared to how much it could hold at that temperature and pressure, expressed as a percentage

Ridge Elongated area of high pressure extending from the centre of a high pressure region (opposite of a *trough*)

Sea breeze Light winds blowing onshore from any body of water when air over land is warmer than air over water, opposite of *land breeze*

Sea smoke Fog caused by cold air flowing over warm water

Shower Intermittent rain or snow of short duration in small or large amounts

Snow Precipitation made up of white or translucent ice crystals, usually 6-sided, often clumped into flakes of various sizes

Stratosphere Layer of the atmosphere from about 11-48km/7-30mi up

Sun dogs Ovoid splotches of light on one or both sides of the sun caused by ice crystals

Sunspots Large cooler regions that appear on the sun's surface as dark spots and which increase in abundance every 11 earth years, heightening displays of aurora borealis and disrupting telecommunications

Supercooled water Water that has cooled below freezing without forming ice crystals

Thunderstorm Local storm, usually from a cumulonimbus cloud, with thunder and lightning but not necessarily rain or snow

Trough Elongated area of low pressure extending from centre of a low pressure region (opposite of a *ridge*)

Tornado (Twister) Strong, rotating column of air reaching from the base of a cumulonimbus cloud to the ground

Troposphere Lower layer of the atmosphere, up to 11-13km/7-8mi above the earth

Virga Streaks of falling rain that evaporate before reaching the ground

Watches & Warnings Advance notices by Environment Canada of coming severe weather; if a watch is issued in your area, maintain your normal routine but watch the sky and tune in to forecasts; if a warning is issued, be alert and prepare

Waterspout A tornado over water

Weather Local short-term regional manifestations of climate

Wind chill factor Effect of wind removing heat from the body

Warm front A warm air/cool air boundary with the warm air advancing over the cool